Computational Fluid Dynamics

an Overview of Methods

by D. James Benton

Preface

Fluid flow is a very broad and complicated topic because of the many different aspects and types of applications. I have covered inviscid flow, which occurs when the viscosity is negligible or the velocities are very large, in two previous texts: *Differential Equations* and *Numerical Calculus*. I may at some point cover creeping flow, which occurs when the viscosity dominates momentum or the velocities are very small. In this text we will cover the range between these two extremes: when viscosity and momentum are both important, at least in some significant part of the flow field.

This is by no means an exhaustive reference on the subject of computational fluid dynamics; rather, it is an introduction and overview, going just far enough to get the reader started along this path with some helpful direction based on years of experience. I have striven to make this a clear presentation, particularly of finite elements, avoiding the traditional esoteric derivations that never seem to arrive at any useful destination. The impetus to undertake this project has arisen from my interacting with many graduate students on Research Gate, who are pursuing CFD and struggling with traditional presentations. You must make decisions when you arrive at a fork in the proverbial road (such as FDM, FVM, or FEM) and I hope this text will provide you with enough information to do that without me imposing my personal preference.

All of the examples in this text are written either in the C language or FORTRAN. CFD codes continue to be written in FORTRAN, even long after the language has fallen into disuse in most every other field of programming except complex variables. It is often said that FORTRAN is more efficient—even faster—for such applications, but I have found this to be more a result of inefficient C than a benefit of FORTRAN. Ultimately, the processor executes machine language instructions, which become the limiting factor with efficiently written code. The C software presented here is ANSI standard and uses no compiler or operating system specific features, except for being compatible with the Microsoft® C compiler. When thus compiled, these examples will run on *any* version of Windows®, including: 95, 98, ME, 2K, XP, Vista, 7, 8, 9, and 10 (32-bit and 64-bit). This is also true of the modified FORTRAN codes included in the archive, which can be found at the link below.

All of the examples contained in this book,
(as well as a lot of free programs) are available at…
https://www.dudleybenton.altervista.org/software/index.html

Navier-Stokes Equations on Structured Grid

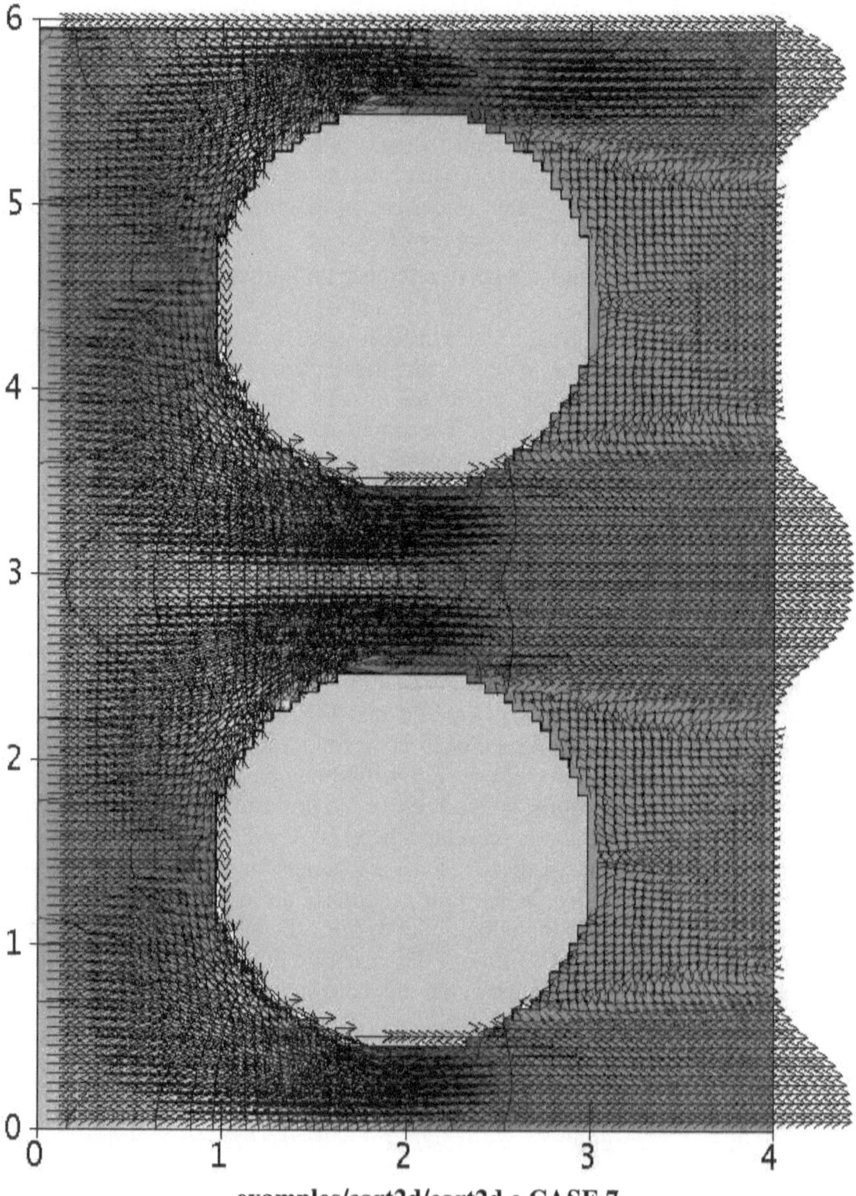

examples/cart2d/cart2d.c CASE 7

Table of Contents　　　　　　　　　　　page

Preface i
Chapter 1. Navier-Stokes Equation.. 1
Chapter 2. Vorticity-Stream Function Method 5
Chapter 3. Finite Difference Method 23
Chapter 4. Finite Volume Method 37
Chapter 5. Finite Element Method... 53
Chapter 6. Turbulence.. 73
Chapter 7. Three Dimensional Flow 85
Appendix A. Tecplot™ vs. TP2... 87
Appendix B. Compilers 91
Appendix C: Finite Differences 95
Appendix D. Arrays.. 97
Appendix E. Green's Lemma 99
Appendix F. Elements... 101
Appendix G: Exact Solutions... 103
Appendix H. Transient Solutions... 105

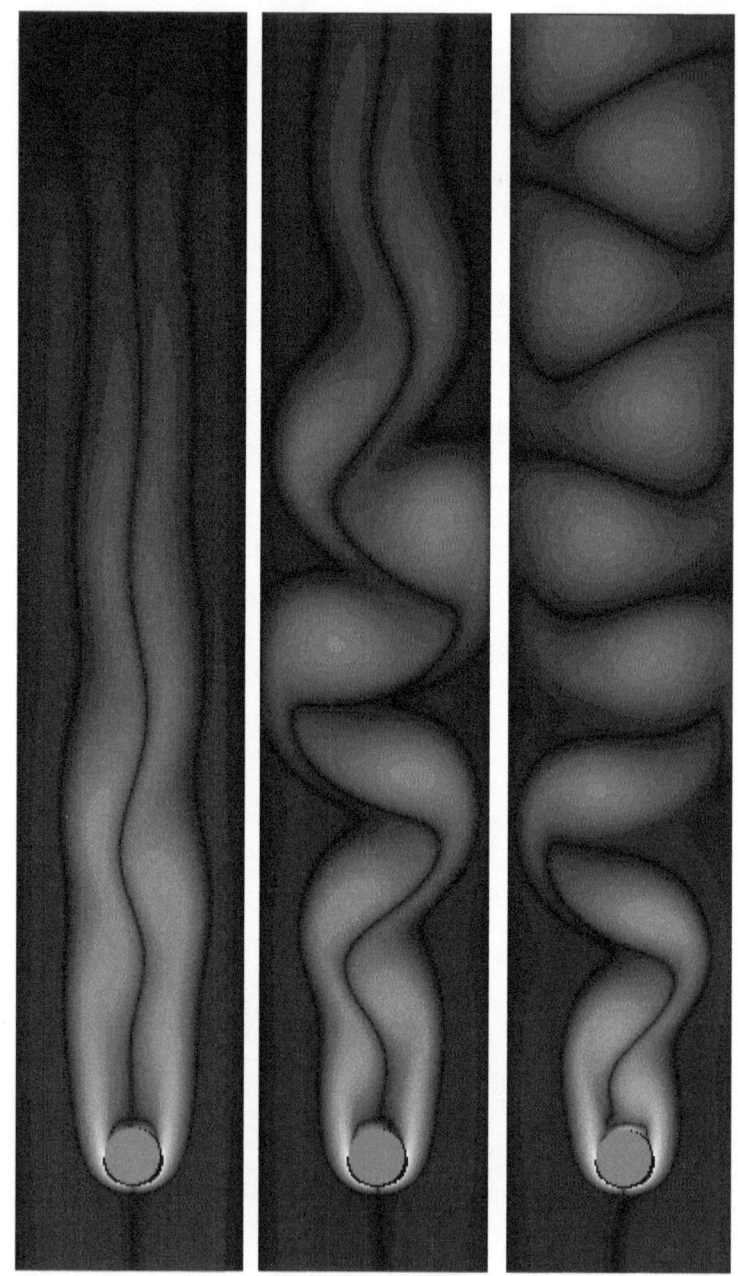

examples/nast2d

Chapter 1. Navier-Stokes Equation

The partial differential equation governing fluid flow is attributed to Navier[1] and Stokes.[2] Their equation is a force balance based on Newton's Second Law, **F=ma**. The sum of all forces on an object is equal to the mass times the acceleration. Before we discuss the fluid terms, we must first take a closer look at Newton's expression. Consider a rocket. Thrust comes from burning fuel and expelling hot gases from the engine. The mass is constantly changing, as the fuel burns. This simple formula is insufficient. Rather, we need to consider the following:

$$\vec{F} = \frac{d(m\vec{V})}{dt} \tag{1.1}$$

In Equation 1.1, **F** and **V** are vectors and **m** is a scalar. Carrying out the differentiation with respect to time, we get:

$$\vec{F} = \vec{V}\frac{dm}{dt} + m\frac{d\vec{V}}{dt} \tag{1.2}$$

$$\frac{d\vec{V}}{dt} = \vec{a} \tag{1.3}$$

Acceleration, **a**, is also a vector. Equation 1.1 is the *conservative* form of **F=ma** in that it considers the conservation of mass. Along with the distinction between *viscid* and *inviscid* flow, which considers or ignores viscosity, μ, there is also the distinction between constant and variable density, ρ. As you peruse the literature on fluid flow, you will also see a distinction between *compressible* and *incompressible*, but this generally refers to velocities near or above the speed of sound (i.e., Mach speeds). While some authors use the term *incompressible* to mean constant density and not specifically low Mach numbers, we will not use the term in that way here to avoid ambiguity.

As Equations 1.1 through 1.3 are vector-based, as long as our coordinate systems (e.g., Cartesian, cylindrical, spherical) are orthogonal, the equations directly apply along each of the coordinates. That is, a force in the **x** direction produces acceleration in the **x** direction, not the **y** or **z** directions. For three dimensions, we will have three conservation equations. Most often, the three Cartesian components of the vector velocity are given the symbols **u**, **v**, and **w** for **x**, **y**, and **z**, respectively.

We will only consider a continuum in this text. We will not be discussing liquid/gas or liquid/solid interfaces except as boundary conditions. We will not

---

[1] Claude-Louis Navier (1785–1836) French engineer and physicist a pioneer in the field of mechanics.
[2] George Gabriel Stokes (1819–1903) Irish physicist and mathematician.

be discussing aerosols, slurries, or fluidized beds. We will also limit our discussion to Newtonian fluids; that is, fluids whose viscous stresses are linearly proportional to the strain. The constant of proportionality is the viscosity. We will not be considering non-Newtonian fluids; that is, fluids that can sustain a shear without flowing (e.g., peanut butter, mayonnaise, glue). We will begin with the conservation of mass in Cartesian coordinates; that is, the *continuity equation*.

$$\frac{\partial \rho}{\partial t} + \frac{\partial(\rho u)}{\partial x} + \frac{\partial(\rho v)}{\partial y} + \frac{\partial(\rho w)}{\partial z} = 0 \tag{1.4}$$

The product terms expand to:

$$\frac{\partial \rho}{\partial t} + \left(u\frac{\partial \rho}{\partial x} + v\frac{\partial \rho}{\partial y} + w\frac{\partial \rho}{\partial z} \right) + \rho\left(\frac{\partial u}{\partial x} + \frac{\partial v}{\partial y} + \frac{\partial w}{\partial z} \right) = 0 \tag{1.5}$$

The *momentum equation* has three parts. Assuming gravity is in the $-z$ direction, these are:

$$\rho\left(\frac{\partial u}{\partial t} + u\frac{\partial u}{\partial x} + v\frac{\partial u}{\partial y} + w\frac{\partial u}{\partial z} \right) = -\frac{\partial p}{\partial x} + \mu\left(\frac{\partial^2 u}{\partial x^2} + \frac{\partial^2 u}{\partial y^2} + \frac{\partial^2 u}{\partial z^2} \right) \tag{1.6x}$$

$$\rho\left(\frac{\partial v}{\partial t} + u\frac{\partial v}{\partial x} + v\frac{\partial v}{\partial y} + w\frac{\partial v}{\partial z} \right) = -\frac{\partial p}{\partial y} + \mu\left(\frac{\partial^2 v}{\partial x^2} + \frac{\partial^2 v}{\partial y^2} + \frac{\partial^2 v}{\partial z^2} \right) \tag{1.6y}$$

$$\rho\left(\frac{\partial w}{\partial t} + u\frac{\partial w}{\partial x} + v\frac{\partial w}{\partial y} + w\frac{\partial w}{\partial z} \right) = -\frac{\partial p}{\partial z} + \mu\left(\frac{\partial^2 w}{\partial x^2} + \frac{\partial^2 w}{\partial y^2} + \frac{\partial^2 w}{\partial z^2} \right) - \rho g \tag{1.6z}$$

The last term in this last equation (ignoring all the others) forms the relationship for hydrostatic pressure, namely:

$$\frac{\partial p}{\partial z} = -\rho g \tag{1.7}$$

In cylindrical coordinates, Equation 1.4 becomes:

$$\frac{\partial \rho}{\partial t} + \frac{1}{r}\frac{\partial(\rho r u)}{\partial r} + \frac{1}{r}\frac{\partial(\rho v)}{\partial \theta} + \frac{\partial(\rho w)}{\partial z} = 0 \tag{1.8}$$

In cylindrical coordinates, Equation 1.6 becomes:

$$\rho\left(\frac{\partial u}{\partial t} + u\frac{\partial u}{\partial r} + \frac{v}{r}\frac{\partial u}{\partial \theta} - \frac{v^2}{r} + w\frac{\partial u}{\partial z} \right) = -\frac{\partial p}{\partial r}$$
$$+ \mu\left(\frac{\partial}{\partial r}\left(\frac{1}{r}\frac{\partial(ru)}{\partial r} \right) + \frac{1}{r^2}\frac{\partial^2 u}{\partial \theta^2} - \frac{2}{r^2}\left(\frac{\partial v}{\partial \theta} \right) + \frac{\partial^2 u}{\partial z^2} \right) \tag{1.9r}$$

$$\rho\left(\frac{\partial v}{\partial t}+u\frac{\partial v}{\partial r}+\frac{v}{r}\frac{\partial v}{\partial \theta}+\frac{uv}{r}+w\frac{\partial v}{\partial z}\right)=-\frac{1}{r}\frac{\partial p}{\partial \theta}$$
$$+\mu\left(\frac{\partial}{\partial r}\left(\frac{1}{r}\frac{\partial (rv)}{\partial r}\right)+\frac{1}{r^2}\left(\frac{\partial^2 v}{\partial \theta^2}\right)+\frac{2}{r^2}\left(\frac{\partial u}{\partial \theta}\right)+\frac{\partial^2 v}{\partial z^2}\right) \tag{1.9θ}$$

$$\rho\left(\frac{\partial w}{\partial t}+u\frac{\partial w}{\partial r}+\frac{v}{r}\frac{\partial w}{\partial \theta}+w\frac{\partial w}{\partial z}\right)=-\frac{\partial p}{\partial z}$$
$$+\mu\left(\frac{1}{r}\frac{\partial}{\partial r}\left(r\frac{\partial w}{\partial r}\right)+\frac{1}{r^2}\left(\frac{\partial^2 w}{\partial \theta^2}\right)+\frac{\partial^2 w}{\partial z^2}\right)-\rho g \tag{1.9z}$$

These same equations in spherical coordinated are similar, though of considerably less utility. The full equations are rarely solved. Most often, certain terms are considered negligible and ignored. This is especially true for spherical problems, which we will not be considering. There are many approaches to solving these equations. The fact that at least two, if not three, of them must be solved simultaneously in order to obtain meaningful results. They are also strongly coupled, precluding separate solution, which might be less complicated.

Common subclasses of solutions include: very high speed (though not Mach) flows where viscous effects are negligible, very low speed flows where viscous effects are dominant, flow very near a surface (boundary layer), and very far from any surface (free stream). We will be considering flows that don't strictly fall into any of these categories; that is, more general applications where density, viscosity, pressure, gravity, and velocity are all significant, at least in some part of the flow field.

Reynolds Number

It is often helpful to consider the fluid properties in Equation 1.6 as the dimensionless ratio known as Reynolds number:

$$\text{Re}=\frac{L\rho V}{\mu}=\frac{LV}{v} \tag{1.10}$$

Slow, viscous flows have a very low Reynolds number and fast, nearly inviscid flows have a very high Reynolds number. What constitutes *low* and *high* depends on the geometry and may span many orders of magnitude. How we determine *low* vs. *high* Reynolds numbers for a particular case is the onset of turbulence. Low Reynolds number flows exhibit no discernable turbulence. The flow appears to be layered (Latin *lamina*) in some visualization techniques and so we call this *laminar*. At high Reynolds numbers, the flow may appear chaotic (Latin *tumultuos*). We will refer to this dimensionless quantity throughout the text.

Chapter 2. Vorticity-Stream Function Method

Before we delve into solving the continuity and momentum equation, we first consider an alternative approach, which dodges the problems associated with solving the conservation equations together. Were we to employ a velocity potential, this would preclude solving viscous flows. Instead, we define a stream function, ψ, which is related to the velocity components.

$$u = \frac{\partial \psi}{\partial y} \tag{2.1}$$

$$v = -\frac{\partial \psi}{\partial x} \tag{2.2}$$

The stream function naturally satisfies the continuity equation:

$$\frac{\partial u}{\partial x} + \frac{\partial v}{\partial y} = 0 \tag{2.3}$$

$$\frac{\partial^2 \psi}{\partial x \partial y} - \frac{\partial^2 \psi}{\partial x \partial y} = 0 \tag{2.4}$$

By substitution, we find that the Laplacian of the stream function plus the curl of the velocity is equal to zero:

$$\nabla^2 \psi + \nabla \times \vec{V} = 0 \tag{2.5}$$

The curl of the velocity is called the vorticity and is given the symbol ω. Equation 2.5 is most often written (i.e., Poisson's Equation):

$$\nabla^2 \psi = -\omega \tag{2.6}$$

Expanding, substituting back into Equation 1.6, and rearranging yields an expression for the rate of change of the vorticity:

$$\frac{\partial \omega}{\partial t} = \frac{\partial \psi}{\partial x}\frac{\partial \omega}{\partial y} - \frac{\partial \psi}{\partial y}\frac{\partial \omega}{\partial x} + \frac{\mu}{\rho}\left(\frac{\partial^2 \omega}{\partial x^2} + \frac{\partial^2 \omega}{\partial y^2}\right) \tag{2.7}$$

Equations 2.6 and 2.7 can readily be solved, one after the other, marching through time from some initial conditions to obtain a steady-state solution. The simplest numerical approximation is that of central finite differences for the first and second derivatives. This can be easily implemented for a regular grid and handle simple geometries.

The Cavity Problem

The classical example for this is the cavity problem, as this was one of the first flow solutions accomplished on a digital computer.[3] The two-dimensional domain is a box with a sliding upper lid, which pulls the fluid along, but does

---

[3] Zuk, J. and Renkel, H. E., "Numerical Analysis of Flow and Pressure Fields in an Idealized Spiral-Grooved Pumping Seal," NASA Technical Note D-6183, 1971.

not leak. The other three sides are stationary. The fluid flows within the box and does not slip at the walls. This digital solution was based on earlier work.[4] The domain is illustrated in the following figure:

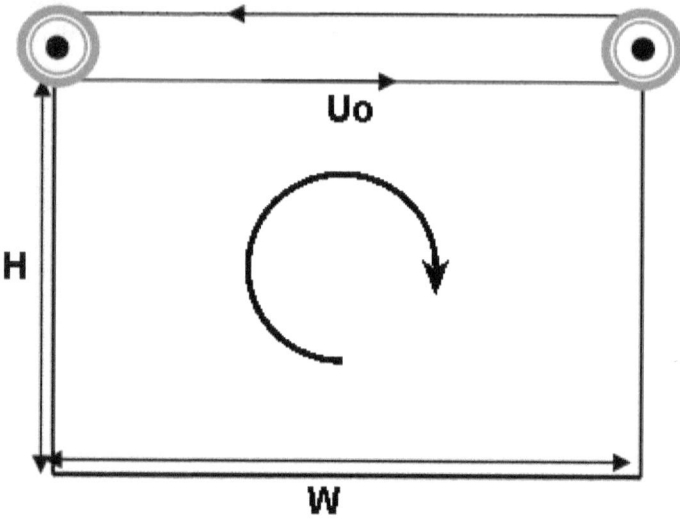

Finite Difference Method

The simplest numerical approximation to these equations is the finite difference method, as illustrated by the following figure:

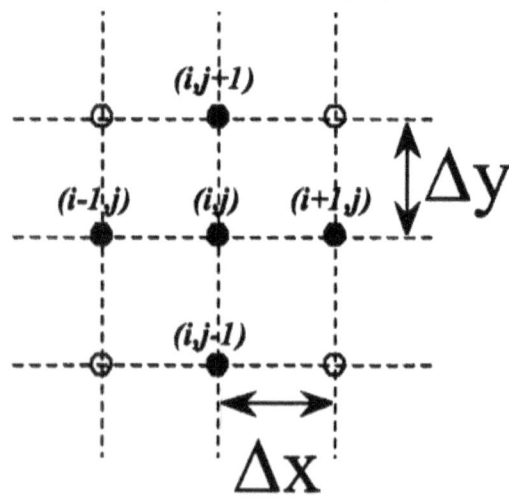

[4] Mills, R. D., "Numerical Solutions of the Viscous Flow Equations for a Class of Closed Flows," Journal of the Royal Aeronautics Society, Vol. 69, No. 658, pp. 714-718, 1965.

Central differences are the first choice, at least for the core. The boundaries require some finesse. The basic relationships are:

$$\frac{\partial \psi}{\partial x} = \frac{\psi_{i+1,j} - \psi_{i-1,j}}{2\Delta x} \tag{2.8}$$

$$\frac{\partial \psi}{\partial y} = \frac{\psi_{i,j+1} - \psi_{i,j-1}}{2\Delta y} \tag{2.9}$$

$$\frac{\partial^2 \psi}{\partial x^2} = \frac{\psi_{i+1,j} - 2\psi_{i,j} + \psi_{i-1,j}}{\Delta x^2} \tag{2.10}$$

$$\frac{\partial^2 \psi}{\partial y^2} = \frac{\psi_{i,j+1} - 2\psi_{i,j} + \psi_{i,j-1}}{\Delta y^2} \tag{2.11}$$

Note that the indices i and j are not always associated with x and y, respectively. See Appendix C for more information on finite differences.

Minimalist Solution

The simplest implementation of Equation 2.6 (i.e., the Laplacian) arises from Equations 2.10 and 2.11. For the homogeneous case (i.e., $\omega=0$) the central value (i,j) is simply equal to the average of the four surrounding ones. The resulting finite difference equation including the vorticity is:

$$\psi_{i,j} = \frac{\Delta y^2 \left(\psi_{i+1,j} + \psi_{i-1,j}\right) + \Delta x^2 \left(\psi_{i,j+1} + \psi_{i,j-1}\right) + \Delta x^2 \Delta y^2 \omega_{i,j}}{2\left(\Delta x^2 + \Delta y^2\right)} \tag{2.12}$$

The simplest time stepping is the fully explicit, forward Euler:

$$\omega_{t+\Delta t} = \omega_t + \Delta t \frac{\partial \omega}{\partial t} \tag{2.13}$$

For some simple cases, we don't have to solve Equation 2.12 accurately at each time step and it isn't strictly necessary that Equation 2.13 yield precise results. If we are only interested in the steady-state solution, start with reasonable initial conditions, and take sufficiently small time steps, the outcome may be adequate—at least for illustration purposes. If the Reynolds number is below 1000, this methodology will work for the cavity problem.

This minimalist solution has been implemented in the vsfm.c code. This and other codes and files related to the cavity problem can be found in the online archive in folder examples\cavity. This example also provides an illustration of files and graphical representation options. The core calculations are listed below:

```
for(t=0.;t<T;t+=dt)
  {
  for(k=i=0;i<Ny;i++)
    {
    for(j=0;j<Nx;j++,k++)
      {
```

7

```c
if(i!=0&&i!=Ny-1&&j!=0&&j!=Nx-1)
  {
  Sx=(S[k+1]-S[k-1])/(2.*dX);
  Wx=(W[k+1]-W[k-1])/(2.*dX);
  Sy=(S[k+Nx]-S[k-Nx])/(2.*dY);
  Wy=(W[k+Nx]-W[k-Nx])/(2.*dY);
  Wxx=(W[k+1]-2.*W[k]+W[k-1])/(dX*dX);
  Wyy=(W[k+Nx]-2.*W[k]+W[k-Nx])/(dY*dY);
  W[k]+=dt*(Sx*Wy-Sy*Wx+nu*(Wxx+Wyy));
  }
else if(i==0&&j!=0&&j!=Nx-1)  /* bottom */
  W[k]=2.*(S[k]-S[k+Nx])/(dY*dY);
else if(i==Ny-1&&j!=0&&j!=Nx-1)  /* top */
  W[k]=2.*(S[k]-S[k-Nx])/(dY*dY)-2.*Uo/dY;
else if(j==0&&i!=0&&i!=Ny-1)  /* left */
  W[k]=2.*(S[k]-S[k+1])/(dX*dX);
else if(j==Nx-1&&i!=0&&i!=Ny-1)  /* right */
  W[k]=2.*(S[k]-S[k-1])/(dX*dX);
else if(j==0&&i==Ny-1)  /* top left */
  W[k]=W[k+1];
else if(j==Nx-1&&i==Ny-1)  /* top right */
  W[k]=W[k-1];
  }
}
for(k=i=0;i<Ny;i++)
  for(j=0;j<Nx;j++,k++)
    if(i!=0&&i!=Ny-1&&j!=0&&j!=Nx-1)
      S[k]=(2.*dX*dX*dY*dY*W[k]
        +dY*dY*(S[k+1]+S[k-1])
        +dX*dX*(S[k+Nx]+S[k-Nx]))/2./(dX*dX+dY*dY);
```

In this same folder you will find a batch file (_compile.bat) to compile the code using the Microsoft® C (see Appendix B). The output is:

```
solving the cavity problem using
the vorticity-stream function method
Re=10
initializing
time-stepping
done
calculating U
calculating V
bounding psi
writing: psi.tb2
writing: omega.tb2
writing: vsfm.v2d
writing: vsfm.tp2
writing: vsfm.plt
```

After the solution is obtained, five files are created. Four of these are for use with TP2: two surfaces for contours (2D tables, having extension TB2), one for the stream function (psi.tb2) and a second for the vorticity (omega.tb2); the 2D velocity vectors (vsfm.v2d); one layout to display the vectors on top of the contours (vsfm.tp2). The fifth is for use with Tecplot™: a combined data file including stream function, vorticity, and velocity vectors. You will also find a corresponding Tecplot™ layout file in this folder: vsfm.lay. The results, as displayed by TP2 are:

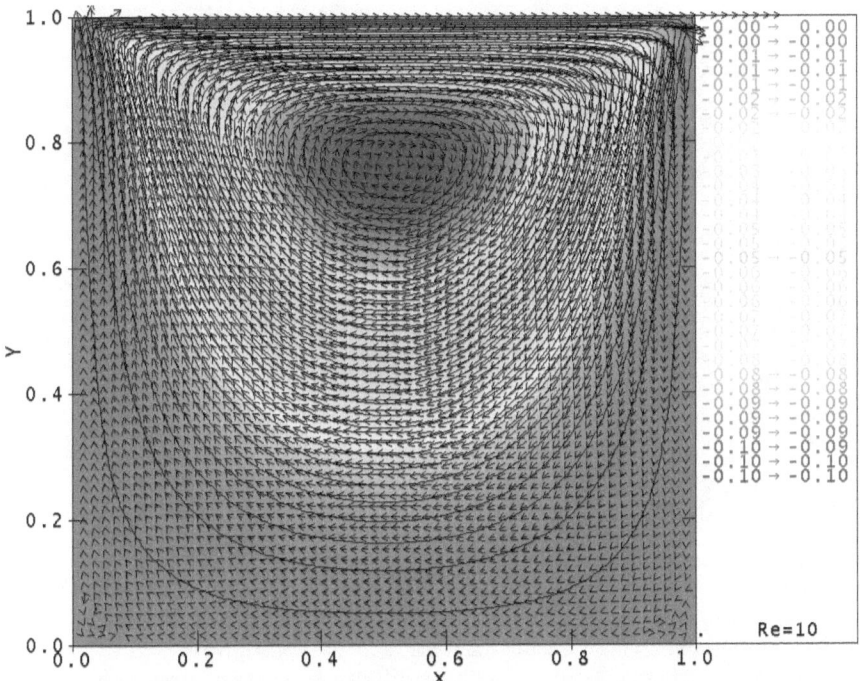

Tecplot™ is similar to TP2 in displaying this type of data. Differences between the two are described in Appendix A. Tecplot™ is an excellent commercial product, which comes with support. TP2 is a freebie with a development and support team of one (me). Both target the Windows™ operating system. For LINUX there is GNUplot, which I know nothing about and with which I can't help you. http://www.gnuplot.info/

The Tecplot™ data structure looks like this:

```
TITLE="Cavity"
VARIABLES="X" "Y" "U" "V" "Psi" "Omega"
ZONE T="results"  I=60, J=60, K=1, F=POINT
DT=(DOUBLE DOUBLE DOUBLE DOUBLE DOUBLE DOUBLE)
0 0 0 0 0 -0
0.0169492 0 0 -0 0 -0.00167194
0.0338983 0 0 -0 0 -0.00595021
0.0508475 0 0 -0 0 -0.00911072
```

The same data displayed by Tecplot™ is:

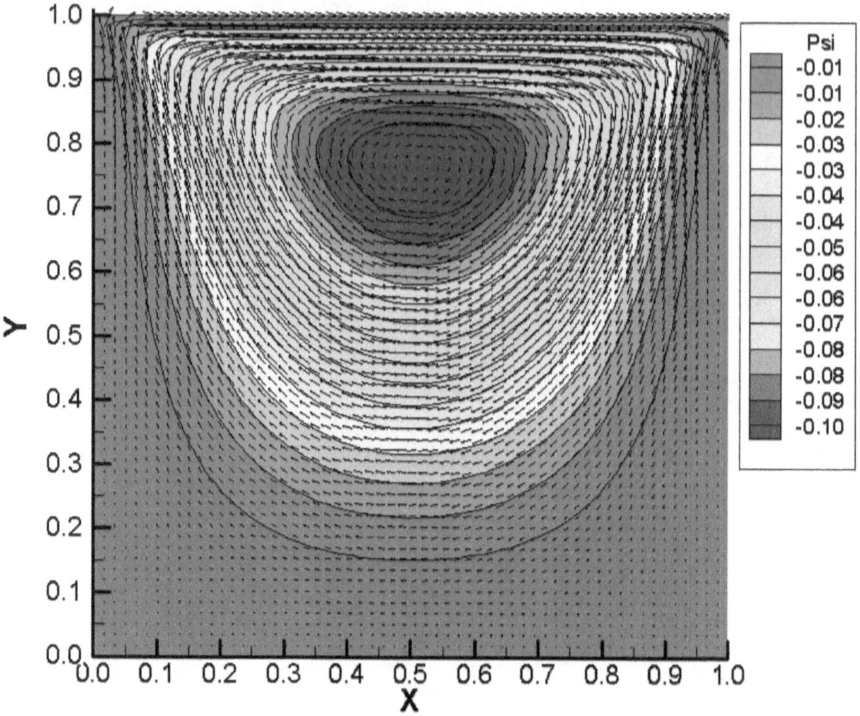

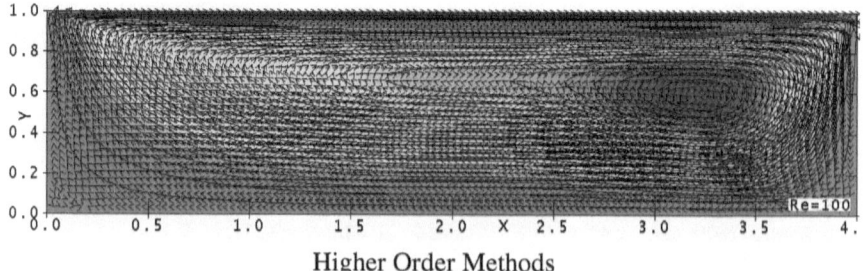

This run is for Re=10 and runs in a few seconds. You can change the kinematic viscosity, v, and lid velocity, Uo, to obtain different values of the Reynolds number. These results are for a 60x60 grid, which you can also change. This particular implementation will also handle x:y aspect ratios other than 1:1, though most presume this to be the case. A 4:1 aspect ratio with Re=100 is shown below:

<u>Higher Order Methods</u>

Leaving the minimalist approach of vsfm.c, we move on to more accurate approaches, specifically ones that solve the Laplacian at each step and may even consider more rigorous time stepping algorithms. A web search for cavity flow turns up a host of articles and pages devoted to this classic problem, many of these reference works from decades ago. Some of the more recent articles reference the work of Erturk, Corke, and Gökçöl (ECG). These authors have specifically addressed the issue of higher order methods.[5,6,7] Erturk has written several FORTRAN codes, which can be downloaded from the following site:

http://www.cavityflow.com/

The associated publications and additional information on the authors can be found at the above site and also on Research Gate:

https://www.researchgate.net/profile/Ercan_Erturk

5 Erturk, E., Corke, T. C. and Gökçöl, O., "Numerical Solutions of 2D Steady Incompressible Driven Cavity Flow at High Reynolds Numbers," Numerical Methods in Fluids, Vol. 48, No. 7, pp. 747-774, 2005.

[6] Erturk, E. and Gökçöl, O., "Fourth-Order Compact Formulation of Navier–Stokes Equations and Driven Cavity Flow at High Reynolds Numbers," *Numerical Methods in Fluids*, Vol. 50, No. 5, pp. 421-436, 2006.

[7] Erturk, E. and Gökçöl, O., "Fourth-Order Compact Formulation of Steady Navier-Stokes Equations on Non-Uniform Grids," *International Journal of Mechanical Engineering and Technology*, Vol. 9, No. 10, pp. 1379-1389, 2018.

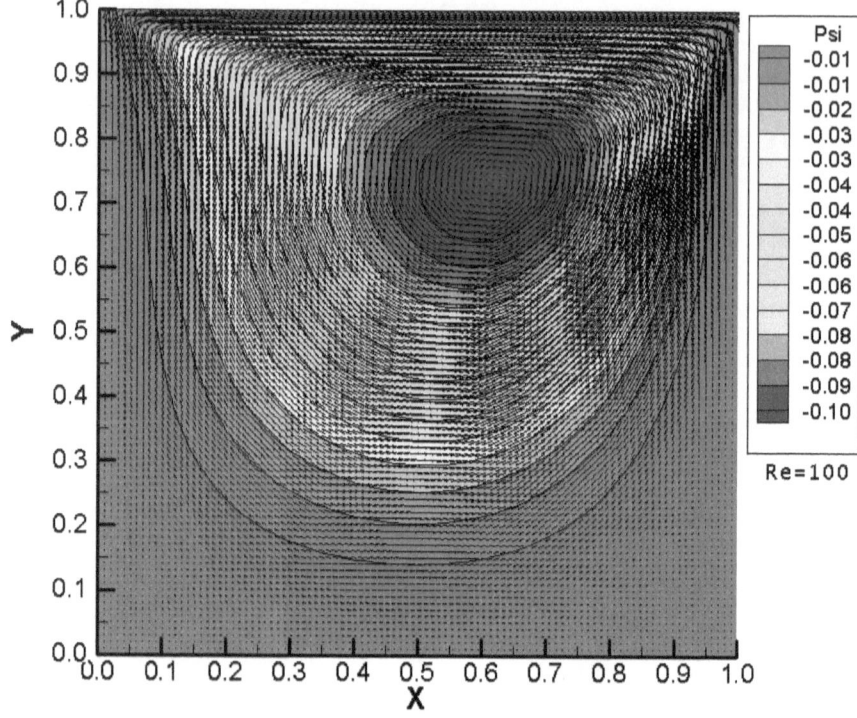

above layout and data: sor2.lay and sor2.plt

I have adapted the vsfm.c code to incorporate the ECG higher order finite difference expressions and Laplacian solver, which uses Successive Over Relaxation (SOR). The second-order code (sor2.c) can be found in the same folder in the online archive. This takes considerably longer to run and is set up for a 1:1 aspect ratio, but yields superior results and works for higher Reynolds numbers. The same five output files are created so that you can readily plot the results, as illustrated on the preceding page. As this higher accuracy solution takes achieve, progress (in estimated percent complete) is displayed during execution. I have also placed a break test within the loop so that if you tap the spacebar (or any key), it will exit the loop and still file the results.

Advanced Finite Differences

The 2nd order finite differences used in the sor2 example here in C and on Erturk's web site in FORTRAN are at the boundaries, not in the core, as the Laplacian operator is already 2nd order. I cover the topic of advanced finite differences in my book, *Numerical Calculus*. I refer to a program called coefs.c, which can be found in the online archive accompanying the text:

https://dudleybenton.altervista.org/software/Numerical_Calculus_examples.zip

In Appendix A of that text, I describe the program in some detail. It will provide the coefficients for interpolation or extrapolation, differentiation or integration, with or without smoothing, of any order, at any point. This is accomplished using orthogonal polynomials so that at no time is a matrix solved to obtain the coefficients. The inputs are: number of points (≥ 2), point of interest (which can be inside or outside the range 1 to n), degree of differentiation (>0) or integration (-1) or neither (0), and level of smoothing (≥ 0). For example, the 2nd order central difference would be 3,2,2,0, which yields: 1,-2,1 (the coefficients in Equation 2.10). The 2nd order 1st partial derivative at point 1 (i,j) would be: 3,1,1,0, which yields: -1.5,2,-0.5 or:

$$\left. \frac{\partial \psi}{\partial x} \right)_i = \frac{-3\psi_{i,j} + 4\psi_{i+1,j} - \psi_{i+2,j}}{2\Delta x} \tag{2.14}$$

The 3rd order 1st partial derivative at point 1 would be: 4,1,1,0, which yields: -1.83333,3,-1.5,0.333333 or:

$$\left. \frac{\partial \psi}{\partial x} \right)_i = \frac{-11\psi_{i,j} + 18\psi_{i+1,j} - 9\psi_{i+2,j} + 2\psi_{i+3,j}}{6\Delta x} \tag{2.15}$$

The 4th order 1st partial derivative at point 1 would be: 5,1,1,0, which yields: -2.08333,4,-3,1.33333,-0.25 or:

$$\left. \frac{\partial \psi}{\partial x} \right)_i = \frac{-25\psi_{i,j} + 48\psi_{i+1,j} - 36\psi_{i+2,j} + 16\psi_{i+3,j} - 3\psi_{i+4,j}}{12\Delta x} \tag{2.16}$$

You can use this simple program to conveniently determine any such coefficients you might need to achieve whatever order of differentiation is required. The source code is provided, which is also an interesting case study discussed in my book, *Orthogonal Functions*.

4th Order + SOR

Erturk also presents a 4th order version of sor2.f, called sor4.f. In order to achieve 4th order, the core difference equations must also be expanded along with the boundary conditions. The solution is quite similar to the previous so that vsfm.c can readily be adapted to this more accurate formulation. The source code (sor4.c) can be found in the same folder: examples\cavity. The results for Re=1000 are shown below:

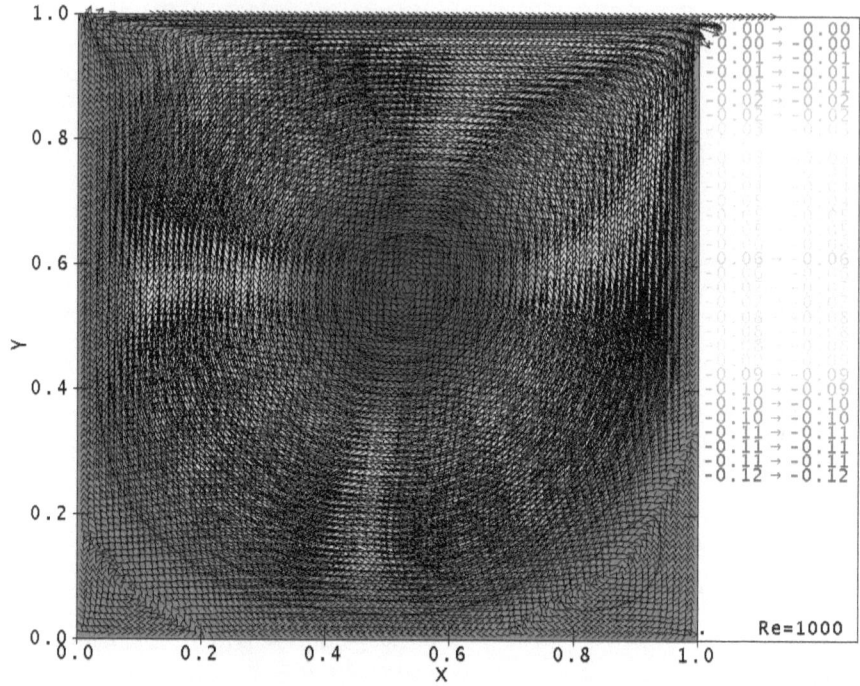

Alternating Direction Implicit (ADI) Method

On the cavityflow web site, Erturk also provides two variants of the Alternating Direction Implicit (ADI) Method for this same problem: a 2nd and 4th order. These are also quite similar to the preceding two so that the vsfm.c code can easily be modified to incorporate these differences. The corresponding codes (adi2.c and adi4.c) can be found in this same folder. The difference equations on the boundary are derived from the work of Stortkuhl, Zenger, and Zimmer.[8] It should be noted that, in spite of the similarity in description (i.e., alternating direction), this is *not* the same as the MacCormack method, which I discuss in my book, *Computer Simulation.*[9]

The differences for the cavity problem are so subtle that no further figures will be included here. Suffice it to say that the codes (adi2.c and adi4.c) work in the same way and create the same files, which can be displayed with either Tecplot™ or TP2. Erturk has two additional FORTRAN programs (erturk2.f and erturk4.f) on the cavityflow site that contain further modifications to this same

[8] Stortkuhl, T., Zenger, C., and Zimmer, S., "An Asymptotic Solution for the Singularity at the Angular Point of the Lid Driven Cavity", *International Journal of Numerical Methods for Heat & Fluid Flow*, Vol. 4, pp. 47-59, 1994.

[9] MacCormack, R. W., "The Effect of Viscosity in Hypervelocity Impact Cratering," American Institute of Aeronautics and Astronautics (AIAA) Paper No. 69-354, 1969.

problem based on his 2005 and 2006 papers previously cited. These are not included in our discussion for the following reason.

Summary Observations

Two things immediately come to mind as we consider these various methods of solving what is a very simple problem with almost trivial boundary conditions: 1) as long as the process converges, it doesn't make a whole lot of difference which method you use and 2) it takes an astonishing amount of time to crunch the numbers. You might ask, "What must a *real* problem take?!" It's no wonder CFD discussions often include parallel processing and cloud computing on a server farm. The runtime is especially noteworthy when you consider that inviscid flow throughout a complicated domain can be solved in seconds using the Boundary Element Method, as I illustrate in *Differential Equations*.

While it is tempting to think that the higher order methods discussed in this chapter can more easily handle large Reynolds numbers (and the titles of some of the articles cited seem to suggest this very thing), just run a few test cases and you will see that this is not the case. So then, what is the advantage of higher order methods? Recall that the first example (sor2) only modified the difference equations at the boundaries. Higher order methods do improve the behavior at the boundaries, especially the corners; but they don't do much of anything for the core, where 2nd order is at least as good as the assumptions implicit in the vorticity-stream function approach.

More Cavity Flows

Before leaving the cavity problem, we should at least consider to simple modifications, which can easily be implemented in vsfm.c by changing the bottom boundary condition. The top boundary moving to the right and the bottom boundary moving to the left produces the circulation pattern shown in the first figure on the following page. The top boundary moving to the right and the bottom boundary also moving to the right produces two counter-rotating cells, as shown in the second figure on the following page. The top boundary moving to the right and the right boundary moving downward produces a single skewed circulation cell, as shown in the first figure on the next page. The top boundary moving to the right and the right boundary moving downward produces two skewed counter-rotating circulation cells, as shown in the second figure on the next page.

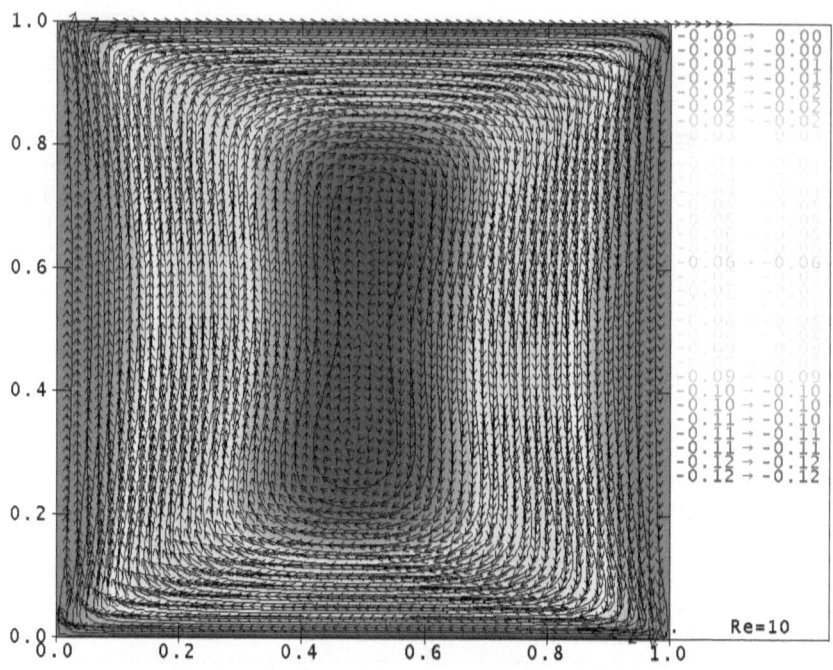

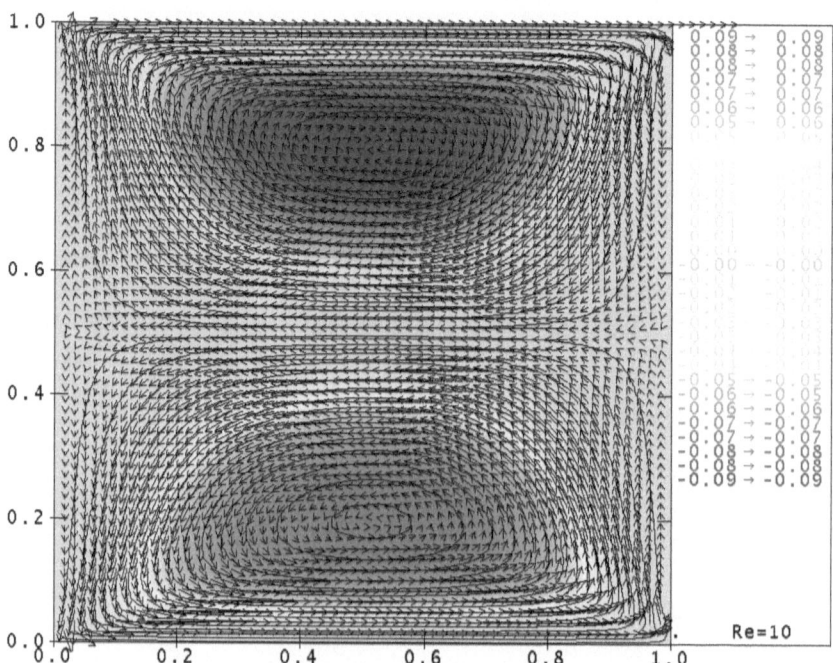

16

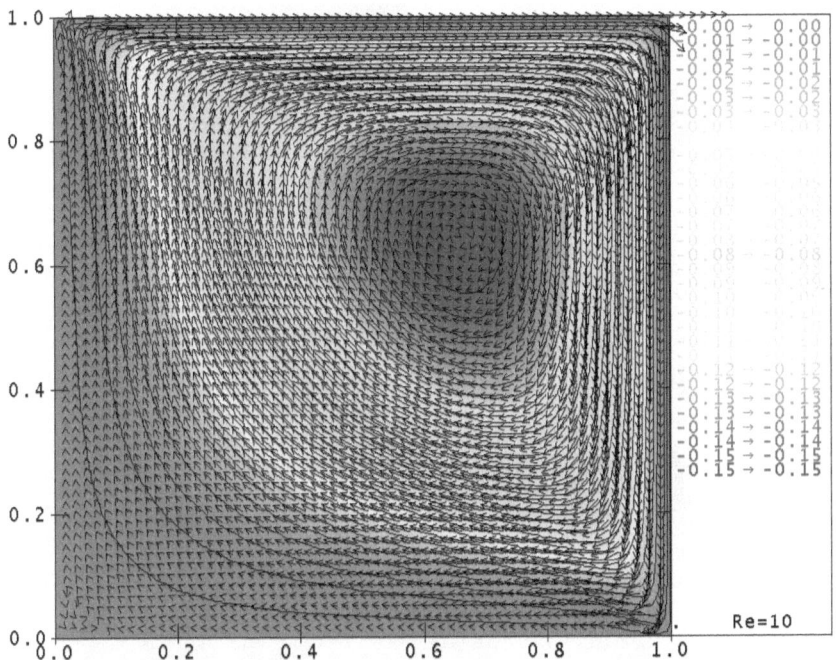

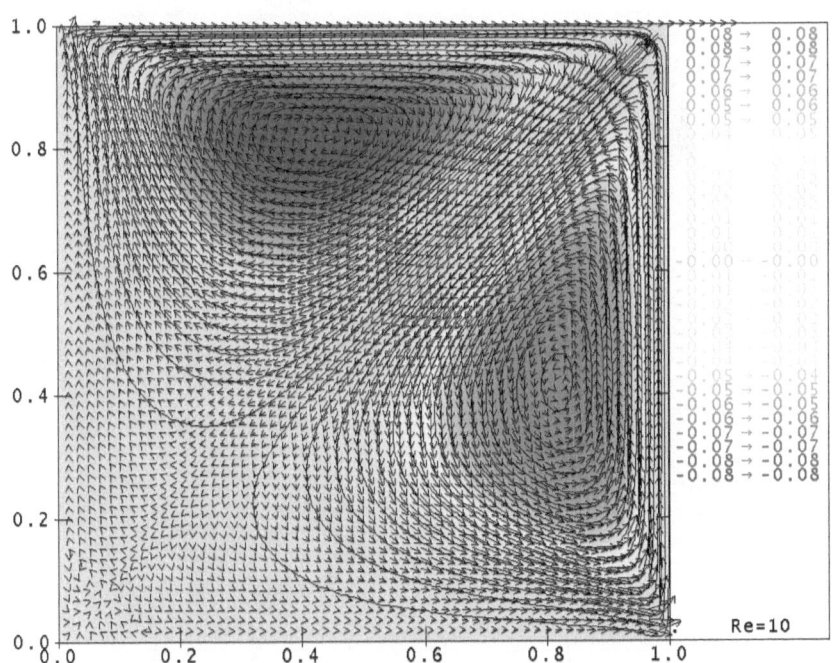

Obstructions

What if we put an obstruction in the cavity? The flow will have to go around it. This is easily accomplished in a few lines of code while initializing the field:

```
for(k=i=0;i<Ny;i++)
  {
  for(j=0;j<Nx;j++,k++)
    {
    W[k]=0.1;
    if(hypot(j-Nx/2.,i-Ny/2.)<=7.)
      {
      S[k]=0.05;
      b[k]=1;
      }
    }
  }
```

The boundary flag, b[], then qualifies Poisson's Equation:

```
for(k=i=0;i<Ny;i++)
  for(j=0;j<Nx;j++,k++)
    if(i!=0&&i!=Ny-1&&j!=0&&j!=Nx-1)
      if(b[k]==0)
        S[k]=(2.*dX*dX*dY*dY*W[k]+dY*dY*(S[k+1]
          +S[k-1])+dX*dX*(S[k+Nx]
          +S[k-Nx]))/2./(dX*dX+dY*dY);
```

The results are shown at the top of the following page. We can move the top and bottom wall and insert an obstruction to obtain flow over a cylinder, which is shown at the bottom of the following page. Then there are four obstructions with bottom and top boundaries moving in the same direction at the top of the next page and four obstructions with bottom and top boundaries moving in opposite directions below that. By now you've seen more than enough flow in a cavity.

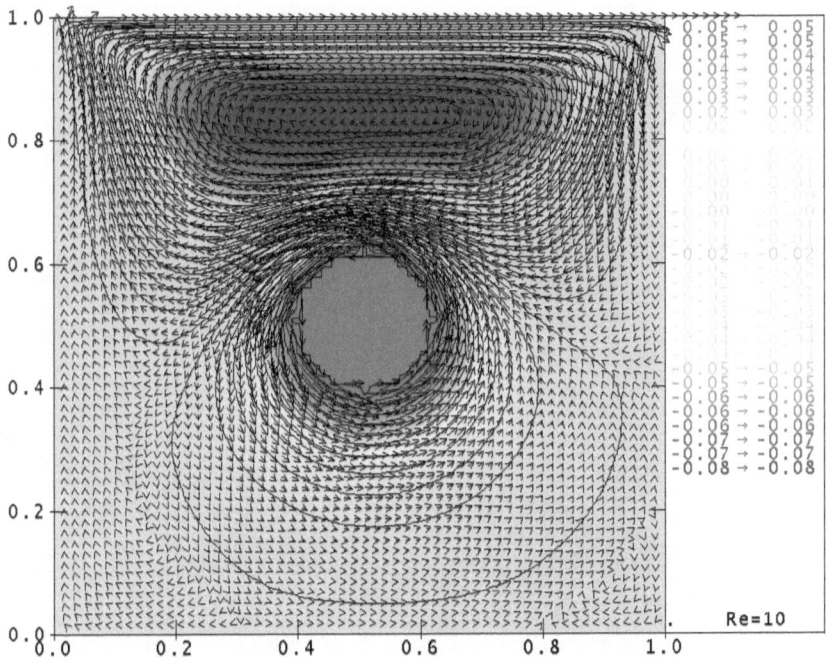

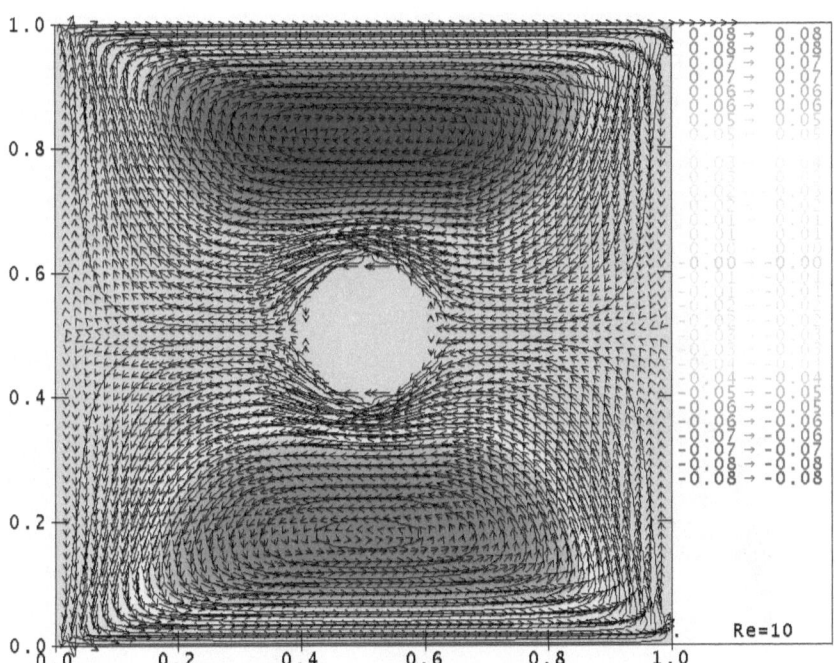

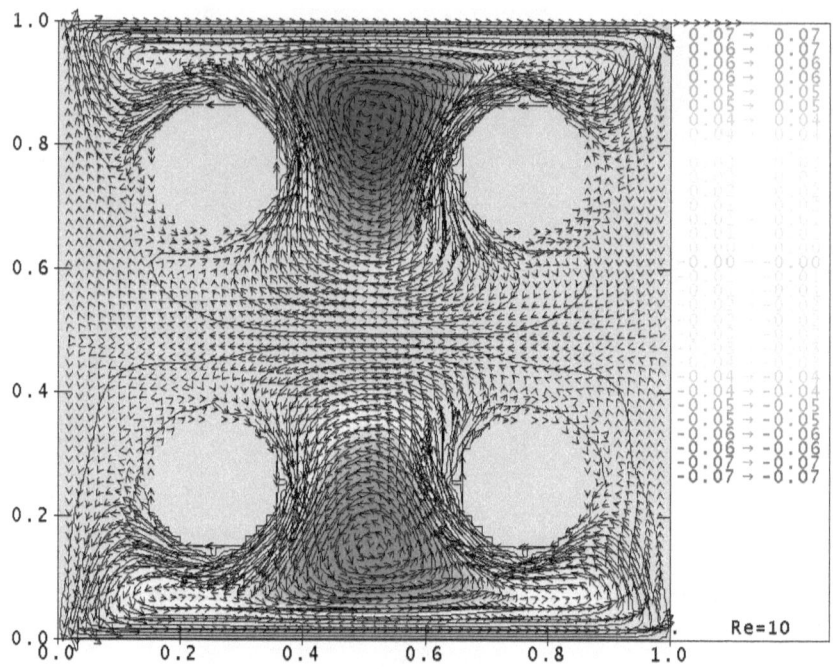

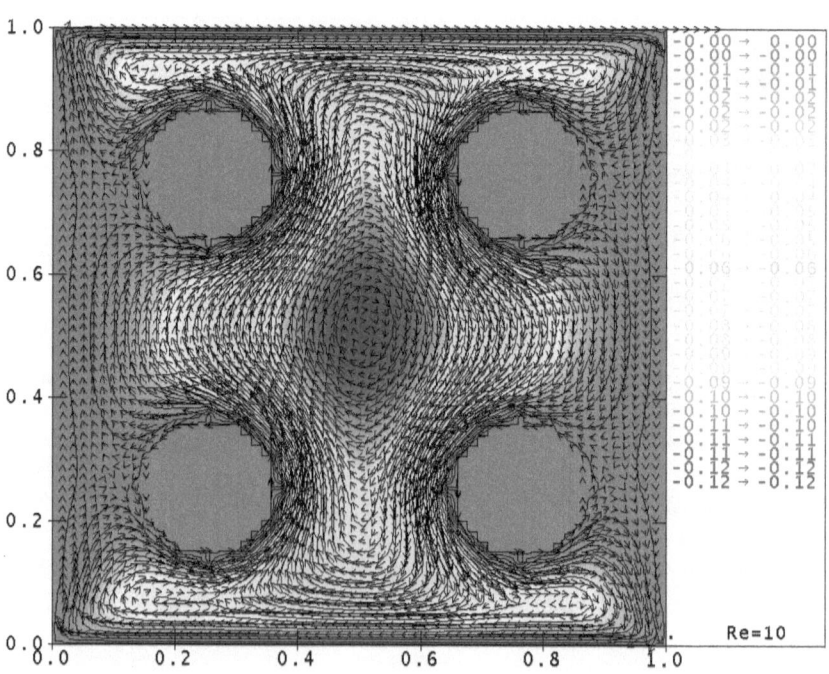

These last few examples illustrate how easily vsfm.c can be adapted to handle somewhat more complicated boundaries. Up until this point, we haven't mentioned how to solve the equations inside and around these obstructions. We know the velocity is zero along the obstructions, which means that the partial derivatives of the stream function must be zero, which means the value of the stream function around the obstructions must be constant, but what value? The stream function inside and around each obstruction must be equal to the surrounding values, which is different, depending on where the obstructions are located. That is, they're not necessarily the same. They appear to be the same in these last two examples only because of the symmetry of the entire cell. This won't be the case if the obstructions aren't symmetric with respect to the flow. In this next example the stream function value around the three obstructions is equal to: Sb=-0.0101598,-0.00895642,-0.0136185.

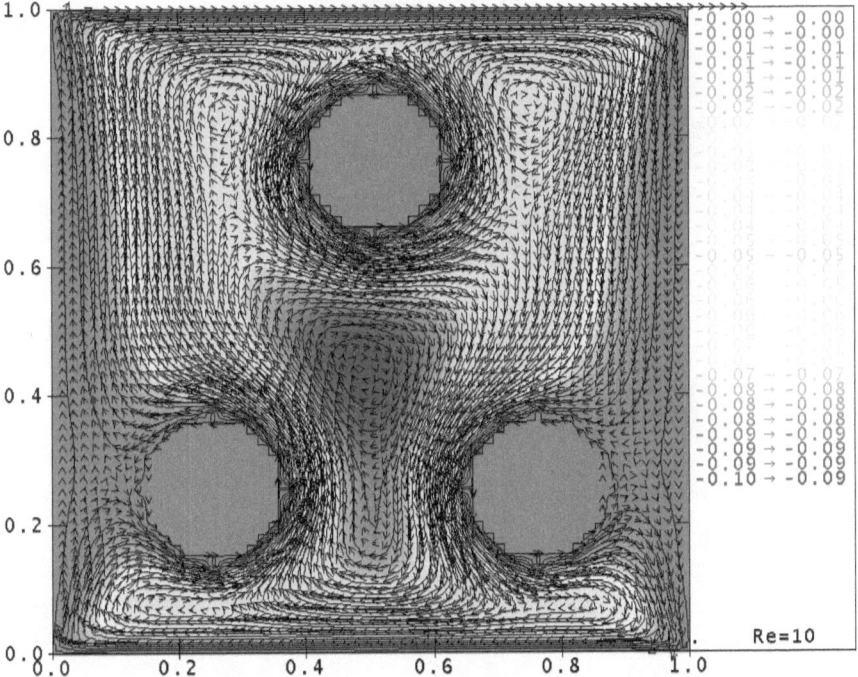

Reversing the direction of the lower boundary produces a different pattern. In this case the three obstruction stream function values are: Sb=0.00644988, 0.00681685, -0.00876505.

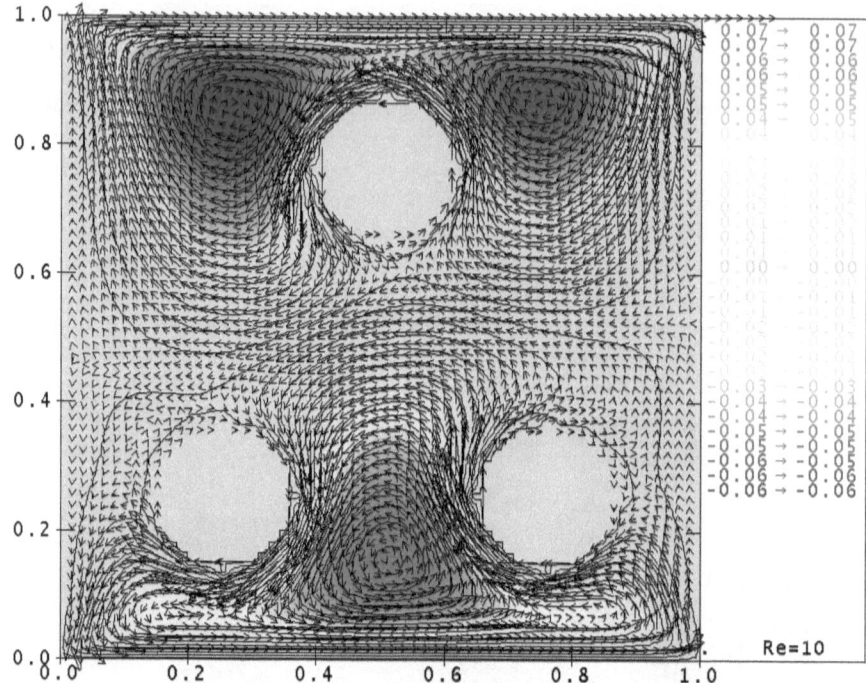

We could continue in this way to construct useful geometries, but what would be the point? Better to move on to finite elements, which naturally fit complex (and useful) boundaries. In the next chapter we leave the vorticity-stream function method and consider the full Navier-Stokes equations for regular grids. After that, we will introduce finite elements.

Chapter 3. Finite Difference Method

The simplest geometry and representation of the conservation of mass and momentum equations presented in Chapter 1 is two-dimensional, Cartesian, structured, finite-difference grids. In this chapter, we will also limit our discussion to constant properties (i.e., density and viscosity). Given these assumptions, Equations 1.4 and 1.5 reduce to:

$$\frac{\partial u}{\partial x} + \frac{\partial v}{\partial y} = 0 \tag{3.1}$$

In this chapter, we will not be considering the vertical dimension, z, which also eliminates gravity so that we will deploy Equation 1.6x and 1.6y in their full form. In order to facilitate solution, we will rearrange these two equations to yield expressions for the rate of change of the two velocity components:

$$\frac{\partial u}{\partial t} = -u\frac{\partial u}{\partial x} - v\frac{\partial u}{\partial y} - \frac{g_c}{\rho}\frac{\partial p}{\partial x} + \frac{\mu}{\rho}\left(\frac{\partial^2 u}{\partial x^2} + \frac{\partial^2 u}{\partial y^2}\right) \tag{3.2x}$$

$$\frac{\partial v}{\partial t} = -u\frac{\partial v}{\partial x} - v\frac{\partial v}{\partial y} - \frac{g_c}{\rho}\frac{\partial p}{\partial y} + \frac{\mu}{\rho}\left(\frac{\partial^2 v}{\partial x^2} + \frac{\partial^2 v}{\partial y^2}\right) \tag{3.2y}$$

Newton's Constant

Before we can solve Equation 3.2, we must first address the units. All of the terms except the ones involving p have dimension length/time². As it appears above, this term has units of force/mass. What is missing is Newton's constant, g_c. For English units, this constant is equal to 32.174 lbm-ft/lbf/sec². For SI units, this constant is equal to 1 kg-m/N/sec².

Pressure Equation

Even with these assumptions, we still have the problem of solving one spatial equation involving both velocity components (Equation 3.1) plus two temporal equations, one for each of the velocity components. This is quite problematic, as there is no direct method for accomplishing this. We have a further problem in that there is no explicit equation for pressure, yet its partial derivatives appear in each of the momentum equations.

A clever way of getting past this obstacle and on to a solution is to consider the ideal gas equation, whether the fluid is a gas or not:

$$\rho = \frac{p}{RT} \tag{3.3}$$

If we substitute Equation 3.3 into Equation 1.5, multiply both sides by RT, and rearrange, we get:

$$\frac{\partial p}{\partial t} = -\left(u\frac{\partial p}{\partial x} + v\frac{\partial p}{\partial y}\right) - p\left(\frac{\partial u}{\partial x} + \frac{\partial v}{\partial y}\right) \tag{3.4}$$

At least for an isothermal flow of an ideal gas, we now have three temporal equations for the three primary variables (u, v, and p), which can be solved using several readily available methods (e.g., finite difference or finite element). Many steady-state solutions are obtained by simply assuming initial conditions, applying constant boundary conditions, and stepping through time until the changes are less than some specified tolerance. The preceding equations can be implemented as finite differences, similar to those in the preceding chapter. The code (cart2d.c) can be found in the online archive in folder examples\cart2d. We will consider the details as we work through several examples. The first is flow to the right through a box with two round obstacles. The velocity vectors and pressure for Re=100 are shown in this first figure:

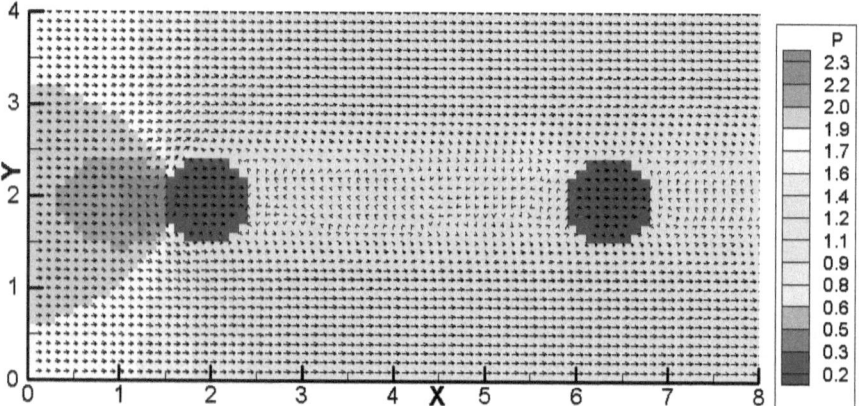

The flow isn't yet fully developed. When it reaches that point, the pressure on the upstream side of both obstructions will be red (high) and on the downstream side will be blue (low). This next figure shows the stream function:

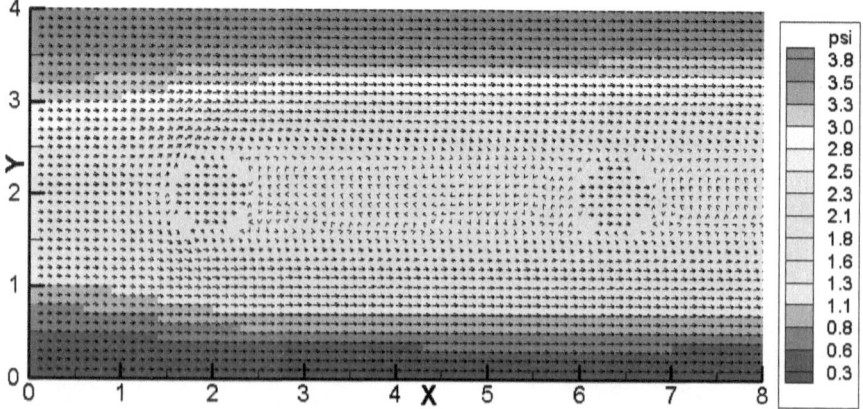

While we don't use the stream function in the solution, it can be easily calculated (4 lines of code) from Equations 2.1 and 2.2 by stepping through the domain after solving for the velocity field:

```
for(j=1;j<=Ny;j++)
  {S[1][j]=S[1][j-1]+U[1][j]*dY;
  for(i=2;i<=Nx;i++)
    S[i][j]=S[i-1][j]-V[i][j]*dX;}
```

We don't use the vorticity in the calculation either, but it can be easily calculated (3 lines of code below the equation) from the curl of the velocity.

$$\omega = \frac{\partial u}{\partial y} - \frac{\partial v}{\partial x} \tag{3.5}$$

```
for(j=1;j<=Ny-1;j++)
  for(i=1;i<=Nx-1;i++)
    W[i][j]=(U[i][j+1]-U[i][j])/dY
            -(V[i+1][j]-V[i][j])/dX;
```

This plot of vorticity makes sense. We know, for instance, that high-speed flow over a cylinder will result in the shedding of vortices like we see in the figure below, which also shows the regular grid.

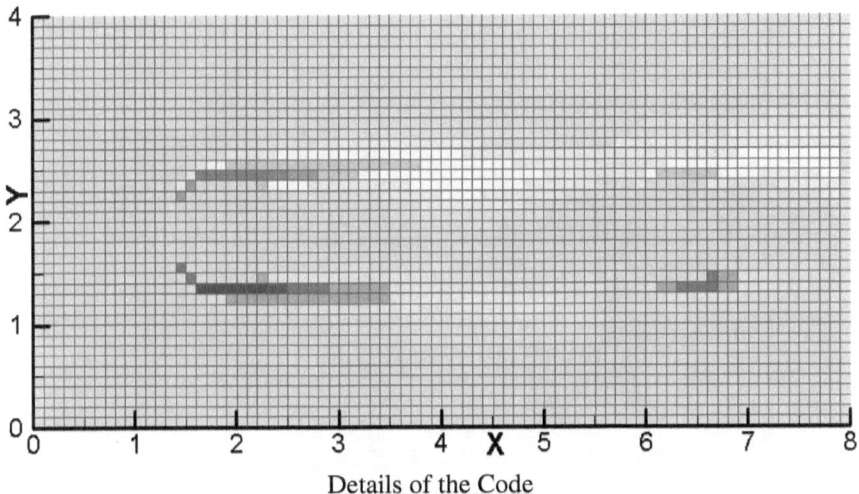

Details of the Code

The code (cart2d.c) creates all the same output files as vsfm.c, so that the results are readily displayed with either Tecplot™ or TP2. The program flow is quite simple. It begins with a few preliminaries, including obstacles and boundaries.

```
Obstacles();
BoundaryFlags();
Initialize();
BoundaryConditions();
```

Several obstacles are built in and selected with conditional compilation statements by setting the variable CASE. The pair of cylinders shown above is

CASE 2. A single cylinder is CASE 1 and a sort of airflow is CASE 3. The solution proceeds with time by first adjusting the time step based on the Courant condition[10], then calculating approximate velocities, then solving Poisson's equation for the pressure field, updating the velocities, and finally applying the boundary conditions.

```
Courant();
ProvisionalVelocity();
PressureEquation();
PoissonSolver();
UpdateVelocity();
BoundaryConditions();
```

After the solution is complete, the vorticity and stream function are calculated and then the output files are created.

```
Vorticity();
StreamFunction();
WriteTB2("psi.tb2",S,0.);
WriteTB2("omega.tb2",W,0.);
WriteV2D("cart2d.v2d");
WriteTP2("cart2d.tp2","p.tb2",P);
WritePLT("cart2d.plt");
```

Flow over the pseudo-airfoil at Re=100 are shown below:

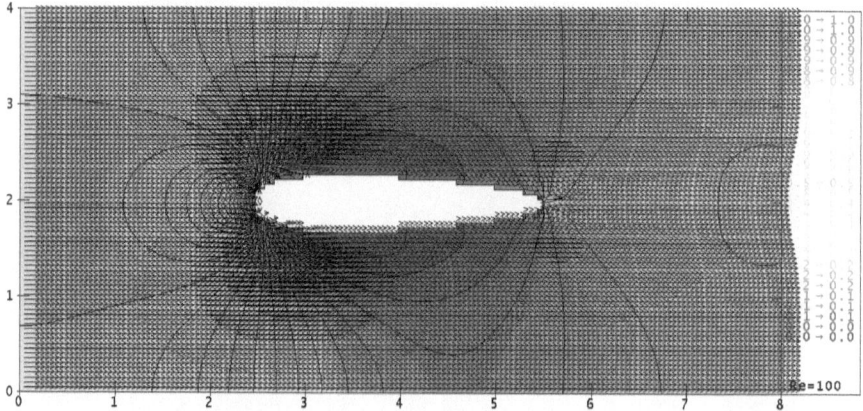

The stream function and vorticity are calculated as before, from the finite difference equations. Although these are not used to arrive at a solution, they are illustrative. The stream function is fairly obvious, but the vorticity field is more interesting.

---

[10] Courant, R., Friedrichs, K., and Lewy, H., "On the Partial Difference Equations of Mathematical Physics", originally published in German in 1928 and reprinted in the IBM Journal of Research and Development, Vol. 11, No. 2, pp. 215–234, 1967.

The stream function is shown in the figure below:

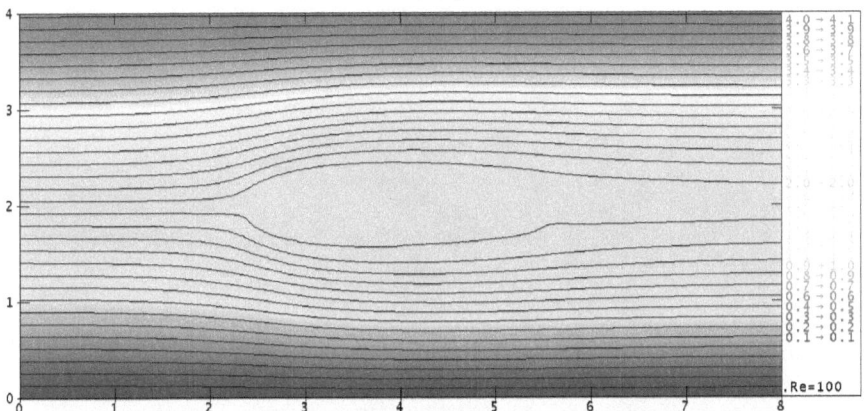

The vorticity is shown in this next figure:

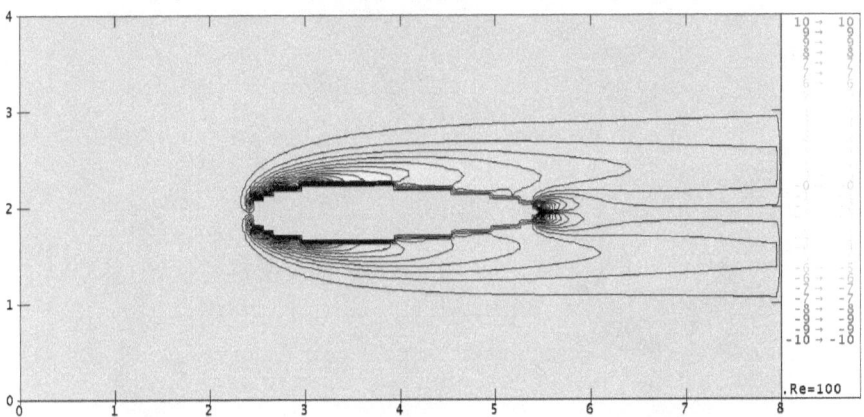

The cart2d.c code is easily modified to handle simple geometries by changing the dimensions (lX and lY), the grid spacing (Nx and Ny), the Reynolds number (Re), and the boundary velocities (Ub and Vb), which are all at the top of the code.

```
#define Nx 80
#define Ny 80
double lX=4.;   /* X length */
double lY=4.;   /* Y length */
double Re=100.; /* Reynolds number */
double Ub=1.;   /* X velocity along boundary */
double Vb=0.;   /* Y velocity along boundary */
```

Obstructions are also easy to insert:

```
#elif CASE==4
  for(i=1;i<=Nx;i++)
```

27

```
{
for(j=1;j<=Ny;j++)
  {
  if(abs(i-Nx/3)<2&&abs(j-Ny/2)<Ny/6)
    B[i][j]=B_O;
  else
    B[i][j]=B_F;
  }
}
```

It's easy to see from this next figure of pressure and velocity why we use airfoils and not flat struts:

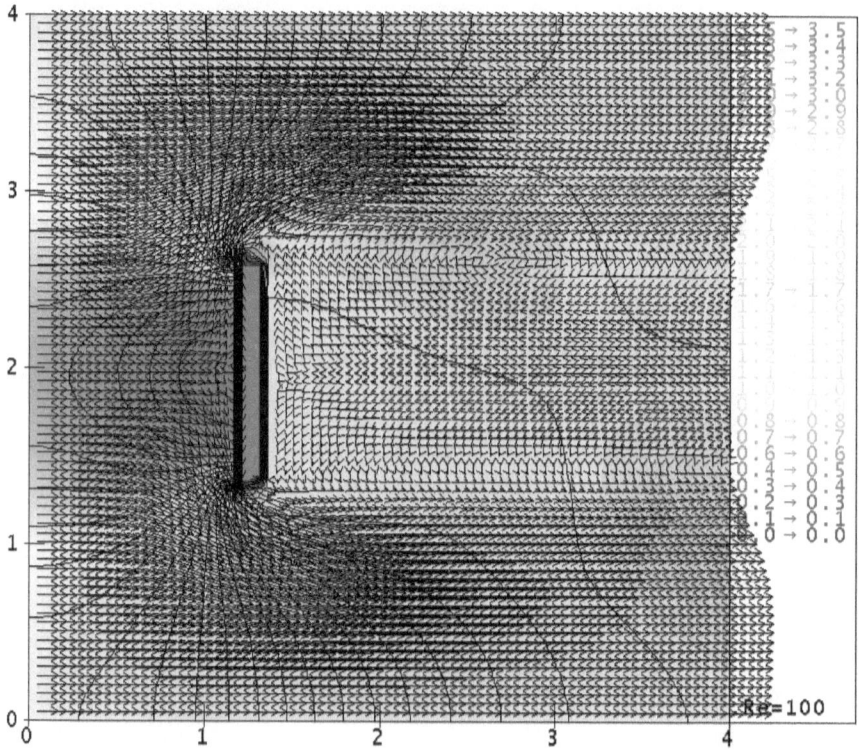

Flow over a perpendicular plate produces a lot of vorticity, as shown in this next figure:

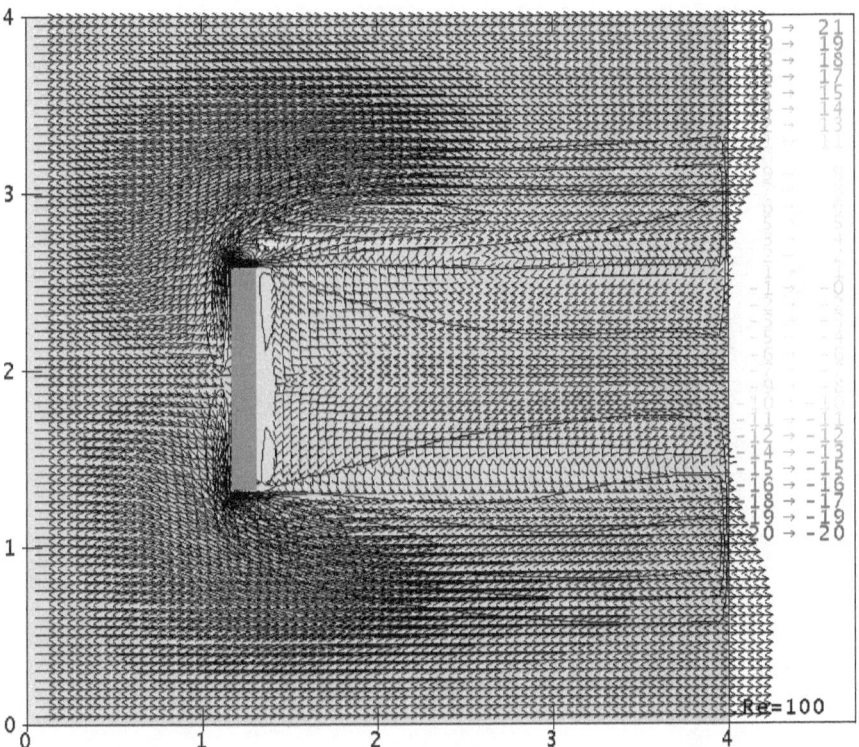

CASE 5 is flow over a wedge, which produces a lot of pressure drop and vorticity. The obstacle is easily entered:

```
#elif CASE==5
  for(i=1;i<=Nx;i++)
    {
    for(j=1;j<=Ny;j++)
      {
      if(i>=Nx/4&&i<=Nx/2&&abs(j-Ny/2)<=abs(i-Nx/4))
        B[i][j]=B_O;
      else
        B[i][j]=B_F;
      }
    }
```

The pressure is shown in this next figure:

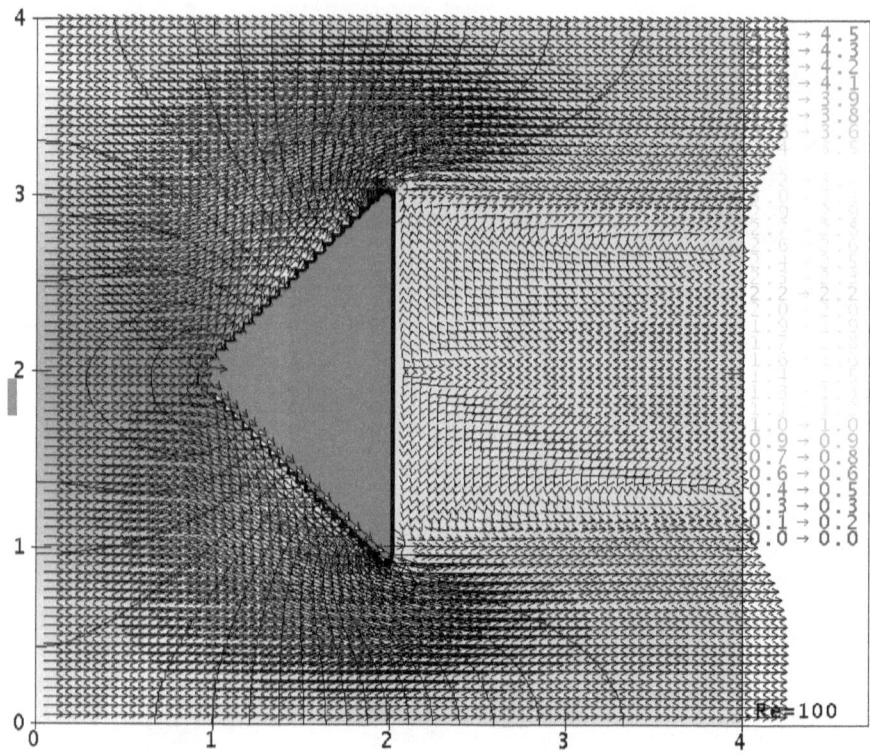

The vorticity is generated at the trailing corners, as shown in the figure below:

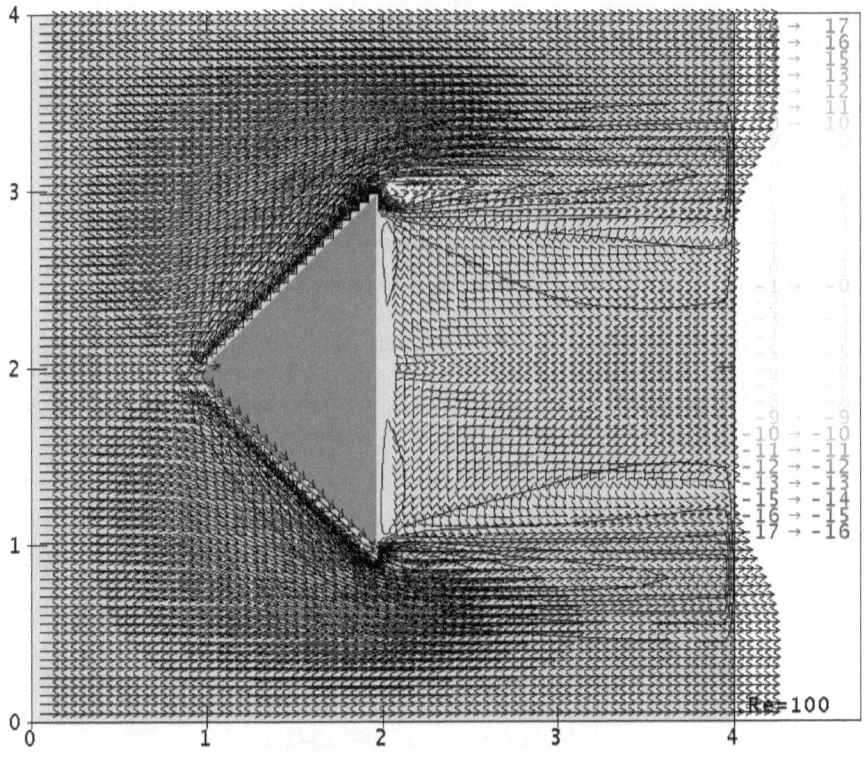

Staggered & Deformed Grids

Before we leave structured grids behind, notice that the equation for pressure (3.4) is first order, while the equations for velocity (3.2) are second order. The two-point finite difference equation for the first derivative is first order on either side and second order in the middle; whereas, the three-point finite difference equation for the second derivative is order two. The finite difference equation for the first derivative straddling the same three points doesn't include the center. This means that, regardless of what the central value is, the first derivative isn't directly impacted, only the second derivative is. This means that, were we to use the simplest finite difference equations to solve this problem, the continuity and momentum equations would be uncoupled at the central point. This approach (which we're not going to implement here) results in what is commonly called the *checkerboard* effect, as illustrated below:

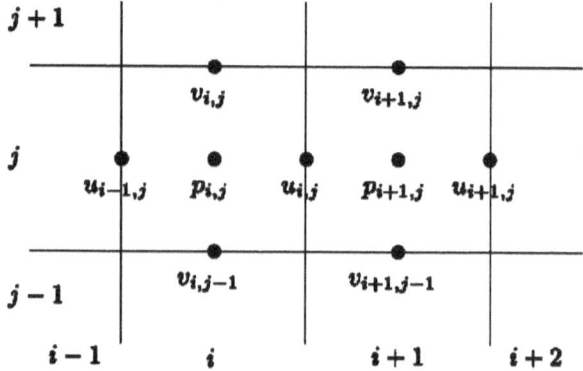

There is no practical way to dampen out or minimize this effect. It inevitably produces undesirable artifacts in the solution. The best approach is to avoid it from the outset by using some other scheme. The most common approach is a staggered grid; that is, solving the pressure and velocities on two different grids that are not at the same points, as shown below:

For a complete domain of (Nx-1)*(Ny-1) cells there will be Nx*(Ny-1) face nodes with associated values of u, (Nx-1)*Ny face nodes with associated values of v, and (Nx-1)*(Ny-1) central nodes with associated values of p. As the nodes with u and v are on the cell faces, the exterior one will all require a boundary condition. None of the central p nodes will be on a boundary; so these won't all require boundary conditions. If there is no boundary condition on any p node, then the solution is ambiguous; thus, at least one p node must be assigned a value. This is a problem when a pressure boundary is critical to the solution.

We don't want velocities on the cell boundaries. When the simulation is complete and we write out the results, we could calculate u and v at the locations of p in the center of each active cell. Without coincident values of u and v, we also couldn't draw velocity vectors, which is the preferred way of visualizing the flow field. Getting around this requires more tricks.

The bottom line is: structured grids are problematic when solving the Navier-Stokes equations, even if the boundaries and obstructions are readily defined in this way. It's simply not worth the effort to solve complex problems using structured grids when unstructured grid (i.e., finite element) methods are available.

Curvilinear and Deformed Grids

Before we leave the discussion of why not to use the FDM, we must consider the fact that boundaries and obstructions are rarely rectangular. If we insist on using the FDM or even the FVM on regular grids, there are undesirable consequences. Consider, for instance, the following domain, which has been studied exhaustively in various laboratories:

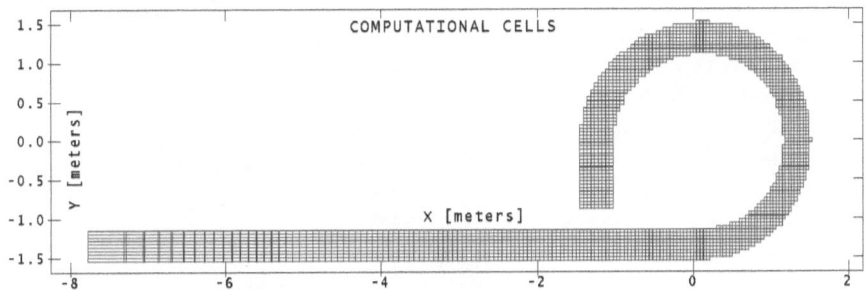

Note the unavoidable stair-stepped boundary...

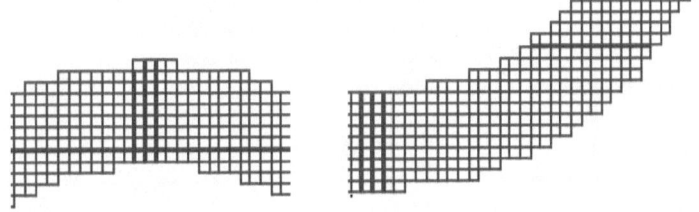

Note the unavoidable artifacts in the velocities where the grid is stair-stepped:

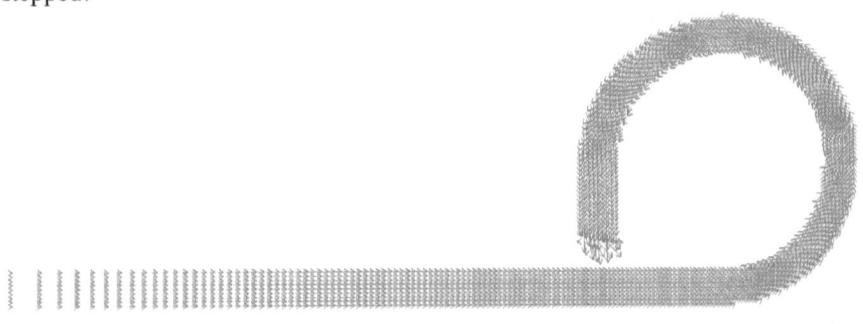

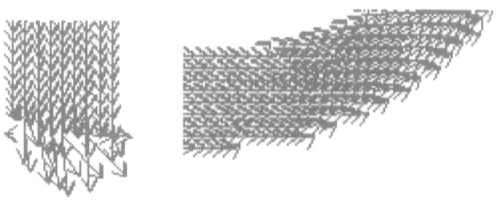

This problem can be lessened by using distorted cells...

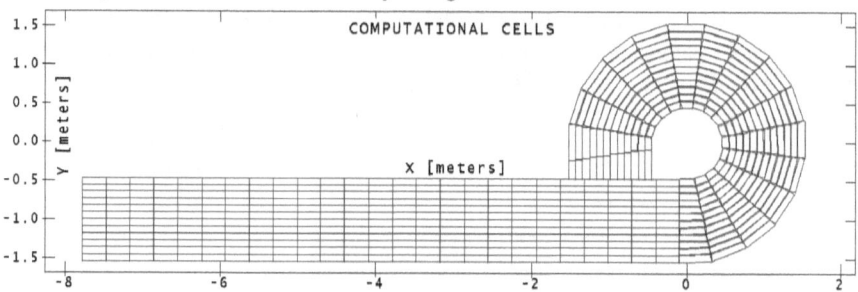

The velocity vectors are somewhat improved with this adjustment:

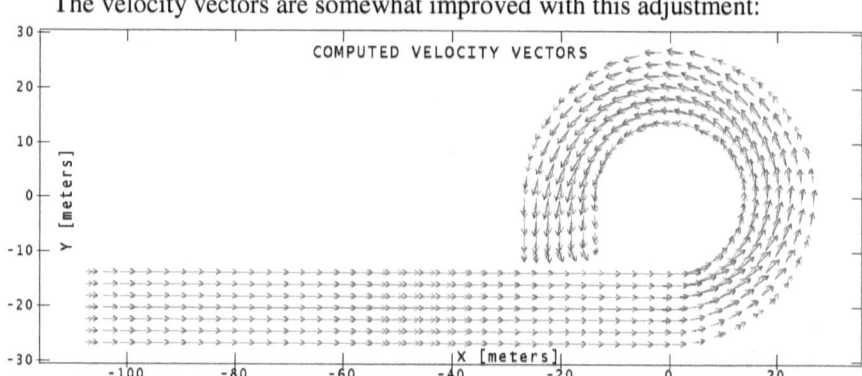

Distorting the elements fixes one problem, but introduces another. We must adjust all of the equations to account for the fact that the elements are rotating, which means we must account for Coriolis effects. First, note the difference in angle, as the fluid is thrust toward the outside of the bend by centripetal force:

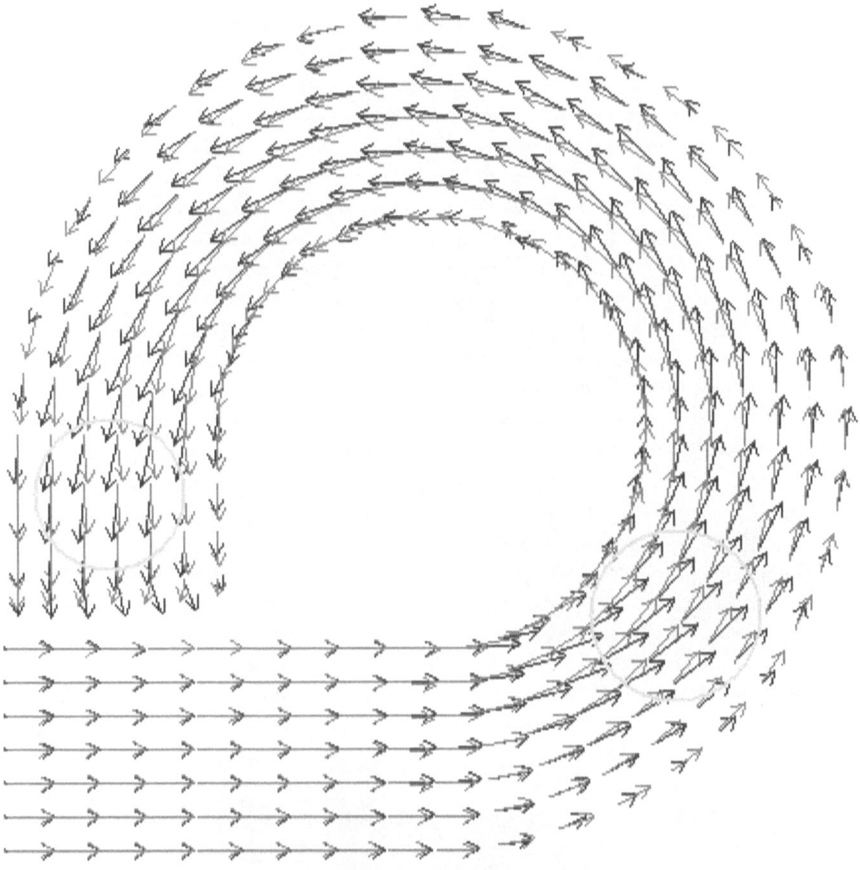

examples\curl\vect.tp2

Also note the discrepancy between the inner and outer velocities when failing to account for curvature of the elements. This is why finite elements are worth the extra effort.

We can still do some interesting problems with FDM and structured grids.

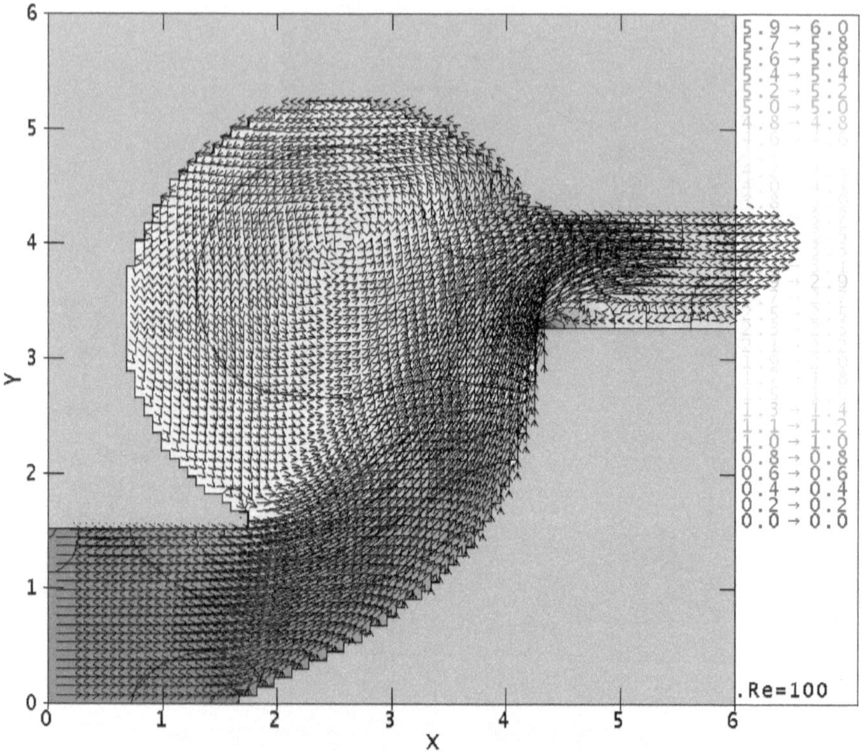

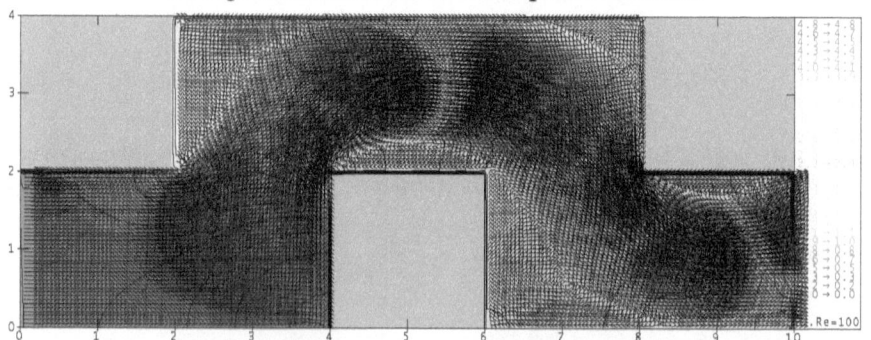

examples\cart2d\cart2d.c with option SCROLL

examples\cart2d\cart2d.c with option DUCT

We clearly need a better approach...

Chapter 4. Finite Volume Method

Structured grids are quite limiting and problematic when trying to solve real-world applications. This gives more than ample motivation to develop alternative methods that will work with unstructured grids, particularly triangular and tetrahedral elements. The Finite Volume Method (FVM) is just such an effort. Based on the control volume formulation, FVM is one of the most versatile and successful discretization techniques used in CFD. It is also the most common, as you will discover by searching the Web.

The cells or elements are considered individual control volumes, in which the variable of interest is located at the center. We first integrate the differential form of the governing equations over each control volume. Basis functions (i.e., interpolators) are used to describe the variation of the key parameters from one cell centroid to the next. These combine to describe a domain over which mass, energy, and momentum are conserved. This strategy can yield adequate results, even for coarse grids—something we don't see with FDM. The approach begins with the following generic expression, where ξ is any conserved property:

$$\frac{\partial \xi}{\partial t} + \nabla \cdot f(\xi) = 0 \qquad (4.1)$$

The function, f, and property, ξ, may represent a vector, such as momentum or a mass flux. The *del* term represents the divergence of the quantity, $f(\xi)$. The conservation equation is integrated over the control volumes, applying the divergence theorem:

$$\frac{\partial}{\partial t} \int_V \xi dV + \oint f(\xi) \cdot \vec{n} dS = 0 \qquad (4.2)$$

where n is the unit outward normal and dS is a differential element on the surface. A variety of methods have been employed to accomplish the integrals, including Gauss quadrature, which gives rise to many variants.

The resulting terms are evaluated as fluxes at the surfaces of each control volume (i.e., element or cell). Because the flux entering one cell is identical to that leaving the adjacent cell, conservation is naturally achieved. The FVM has been used extensively with good success, for instance, the commercial codes Fluent[11] and Phoenics[12] use the FVM, while AutoDesk's CFD code[13] uses the FEM, which we will discuss next. Suhas Patankar was instrumental in the

---

[11] Fluent® is a CFD code marketed by ANSYS®. Here is a link to their wet site
https://www.ansys.com/products/fluids/ansys-fluent
[12] Phoenics® is a CFD code marketed by CHAM®. Here's a link to their web site:
http://www.cham.co.uk/phoenics.php
[13] AutoDesk also markets AutoCAD. Here's a link to their web site:
http://www.cham.co.uk/phoenics.php

development of the FVM and also founding of CHAM (Concentration, Heat, And Momentum). He has authored numerous papers and several texts on the subject.[14]

Open FOAM, the most popular free CFD code, uses the FVM. It has a large user base across most areas of engineering and science, from both commercial and academic organizations. Open FOAM has a wide range of features to solve complex fluid flows involving chemical reactions, turbulence, heat transfer, and more. It has been rigorously evaluated, tested, and validated. The web site is:

https://www.openfoam.com/

We will not rehash the extensive documentation available on the Web. Our purpose in this text is to present the theories behind several approaches and illustrate the results so as to compare and contrast. You can also find many current projects using Open FOAM on Research Gate:

https://www.researchgate.net

We will not present Open FOAM examples here, as the code and associated files are quite complicated—so much so, that the important details of the FVM would be lost in the process. Instead, we will consider a much less complex code that will serve to illustrate the method. There are numerous codes available on the Web. It is disappointing how few of these will run on Windows® without extraordinary measures. Running in a LINUX box on top of Windows is not a practical solution and should not be necessary. Many of the available FORTRAN codes require the Gnu compiler, which is tethered to LINUX.

Blazek's Unstruct2D

One of the few free source codes that employ the FVM and is relatively straight-forward, comes from Jiri Blazek of CFD Consulting & Analysis (http://www.cfd-ca.de/). This program provides solutions of the 2D Euler and Navier-Stokes equations on unstructured, triangular grids. It can handle several fluid modes, including: ideal gas model and laminar flow using viscosity computed by Sutherland's law. Roe's flux-difference splitting scheme is employed and Venkatakrishnan's limiter is also implemented. The explicit multistage time-stepping scheme uses Runge-Kutta, with global and/or local time steps. Preconditioning is also available for low Mach numbers. Central implicit residual smoothing can be specified, along with characteristic boundary conditions for external and internal flows. There is also a special initial solution for compressor and turbine blades.

We will only discuss some of the many options available with Unstruct2D. Perhaps the most important quality of this code for our current purposes is that, after minor modifications, it can be compiled by DEC™ FORTRAN with a single command, from a single source file, in a single folder, without any errors.

---

[14] Patankar, S.V., *Numerical Heat Transfer and Fluid Flow*, Hemisphere, 1980.

This means that the executable will run on *any* of Windows®, including: 95, 98, ME, 2K, XP, Vista, 7, 8, 9, and 10 (32-bit and 64-bit). You can find the modified source code (unstruct2d.f90), executable, five example input and output files, plus further documentation in the online archive in folder examples\ unstruct2d. The program requires no installation, no DLLs, and no particular version of anything. The source code is only 173 kB and the executable is only 389 kB. Versions of the original (unmodified) source code can be found in several locations on the web, including the following links:

https://github.com/nuaawubin/blazek_codes

https://github.com/truongd8593/Unstruct2D

https://github.com/liujiamingustc/Unstruct2D

The unstruct2d code has been modified to write output files that can be readily plotted by either Tecplot™ or TP2 (the former is commercial and the latter is a freebie), also on any version of Windows®. I hope that by making this slightly modified version of the code available that will run on 97% of the worlds' computers, this excellent program will be more widely used, especially by graduate students. Dr. Blazek's book, *Computational Fluid Dynamics: Principles and Applications, 3rd Edition*, covers many topics in much greater detail than is possible in this text.[15]

Example 1: Flow over a Flat Plate

Before we get into any of the details of how the equations are built and solved, implementing the FVM in this case, we will consider perhaps the simplest example: flow over a flat plate. All of the input files (fplate.inp fplate.ugr) and output files (fplate.2dv, fplate.v2d, fplate.plt) can be found in the same folder, along with a batch file to launch the program and rename output files (fplate.bat) and a Tecplot™ layout (fplate.lay). Just double-click on the batch file (fplate.bat) or drop the primary input file (fplate.inp) onto the executable (unstruct2d.exe) to solve the problem.

The domain is a simple rectangle and the grid consists of triangles of similar size. This is not typical for unstructured grids and triangular meshes in particular. The cells (elements) are smaller and closer together, squashed around the vertical line $x=0$. This is transition point where we would expect a boundary layer to change from laminar to turbulent, which is why the grid is created in this way. One test of the model is to see if the velocity profile and pressure gradient changes at this point.

---

[15] https://www.elsevier.com/books/computational-fluid-dynamics/blazek/978-0-08-099995-1

The grid is shown in the following figure:

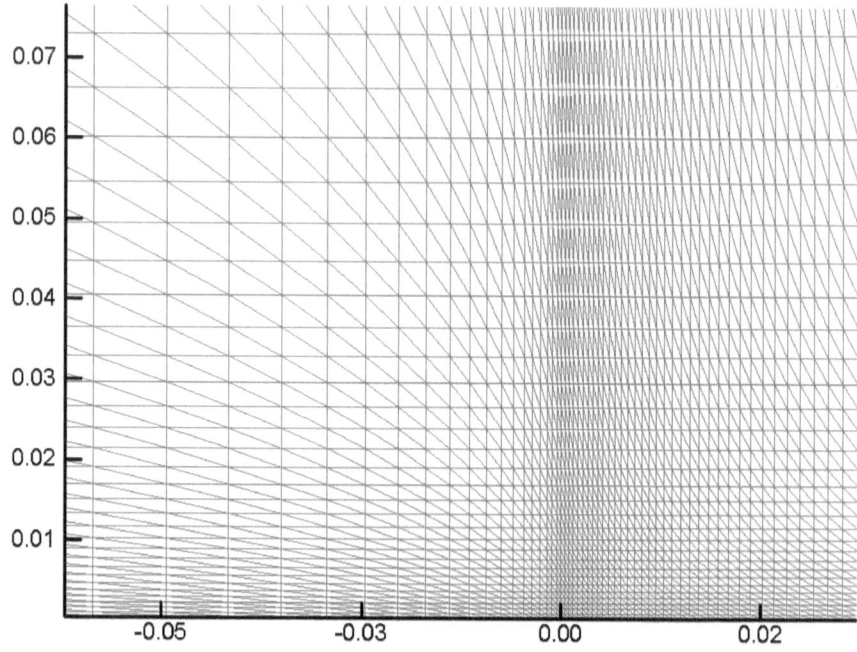

A closer look near *x=0* shows the same structure:

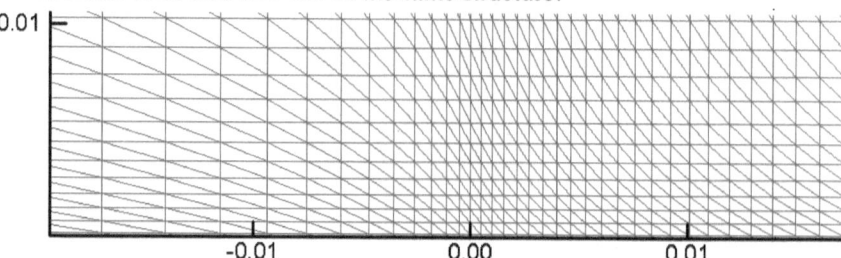

In order to best display the results; we want the pressures at the nodes (element vertices) and the velocity vectors at the center of each element. I have modified the code to output the results in this form. The pressure is a continuous *field*, while the velocity vectors are not—at least conceptually, which is how we want to see them. There are so many elements (and so many velocity vectors) that these lie on top of each other and totally obscure the results. Tecplot™ has a *skip* feature that allows you to draw one out of 5, 6, 7, or however many you want. the layout (fplate.lay) has been adjusted to produce the best graphic.

The pressure field and velocity vectors are shown in this next figure:

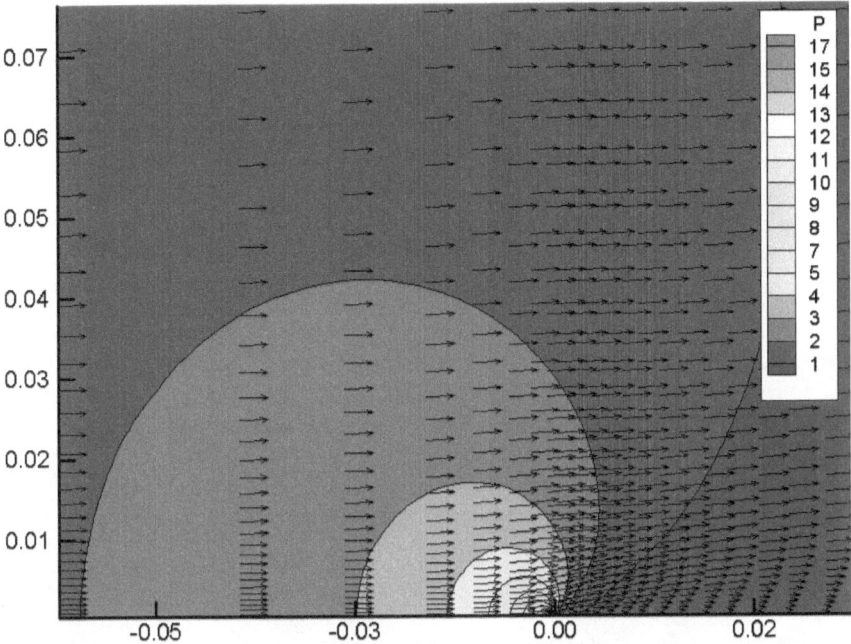

We see that the pressure and velocity profile changes near the transition point, which is a validation of the calculations, including the turbulence model, which we will discuss later. A closer look at the transition point is shown in this next figure:

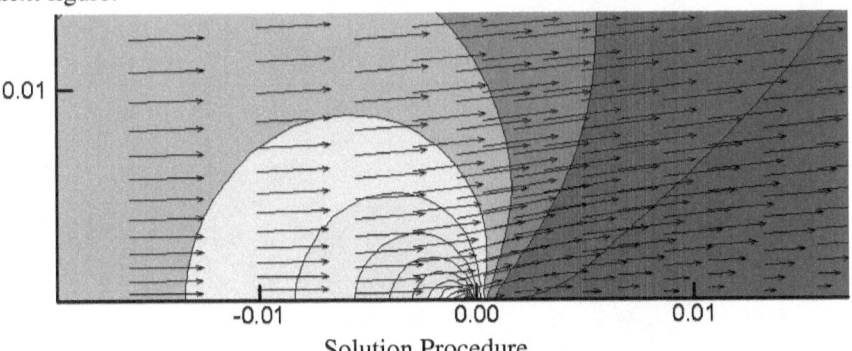

Solution Procedure

Before we examine the next example, we will consider the solution procedure. In this program (unstruct2d.f90), the solution is found in terms of transform variables. For instance, u and v are not conserved, but a combination of ρu and ρv are. The solution will be in array $cv()$. As we write this out to a file for plotting, we first calculate $u=cv(2,i)/rho$ and $v=cv(3,i)/rho$. These are called

41

conservative and *primitive* variables, respectively. We find the solution in terms of conservative variables and write it out in terms of primitive variables. Subroutine Cons2Prim() performs the calculations in one direction and Prim2Cons() performs the opposite. Each group of conservative variables is obtained by the following code inside several loops. The first two lines build the matrices and the third line solves the 5x5 problem: [dmat]=[pmat][gmat1], which must be performed repeatedly:

```
call Cons2Prim(wvec,wpvec,H,q2,theta,rhoT,0D0,hT,gmat1)
call Prim2Cons(wvec,wpvec,H,rhop,rhoT,0D0,hT,pmat)
call MatrixTimesInverse(wpvec,q2,pmat,gmat1,dmat)
```

Example 2: Transonic Flow over a Bump

The second example is high-speed flow over a 10% circular bump. Viscous terms are presumed negligible in the conservation of momentum. This option is set with "E" on line 25 of the input file (channel.inp). Of course, if the flow is inviscid, there are far more efficient ways to solve this problem, including the Boundary Element Method (BEM), which I cover elsewhere. Again, there is a batch file (channel.bat) to facilitate running the model and renaming the output files. I have also provided a Tecplot™ layout (channel.lay). The grid is more interesting for this problem:

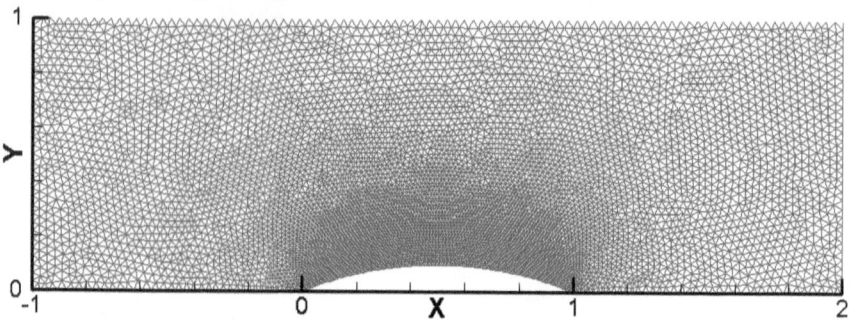

The elements are concentrated near the bump:

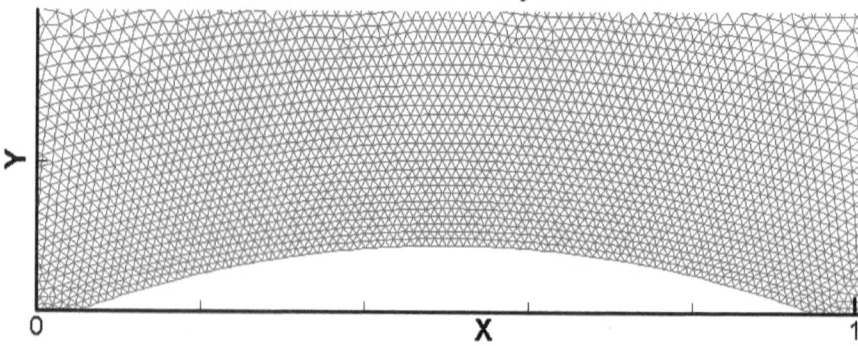

42

The results are shown in the following figure:

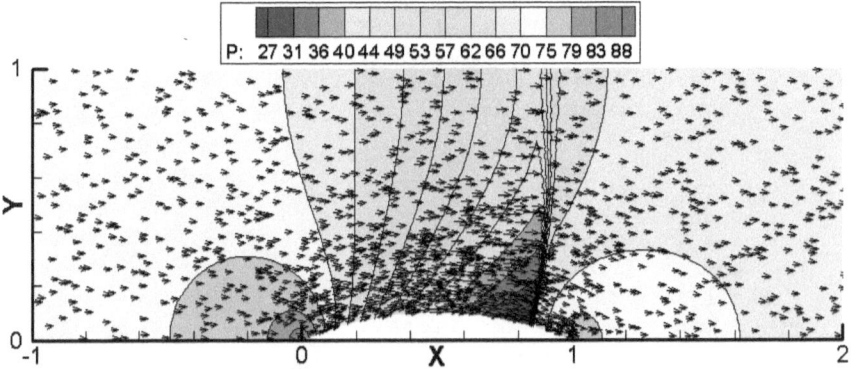

In the vicinity of the bump:

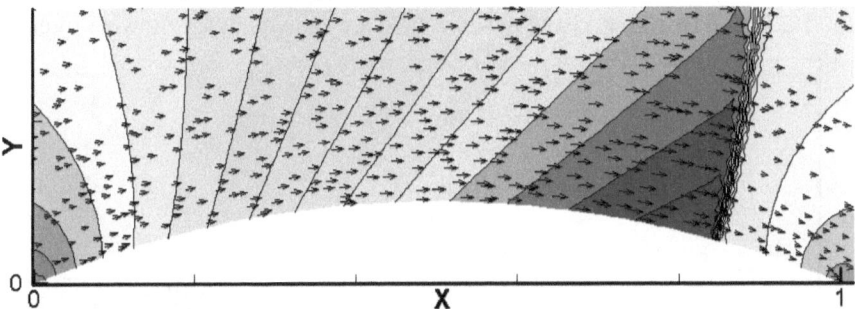

The following section of code is where the calculations differ for these two examples:

```
subroutine DependentVarsAll
  gam1=gamma-1D0
  rgas=gam1*cpgas/gamma
  g1cp=gam1*cpgas
! Euler equations
  if(kequs.eq."E")then
    do i=1,nnodes
      rhoq=cv(2,i)*cv(2,i)+cv(3,i)*cv(3,i)
      dv(1,i)=gam1*(cv(4,i)-0.5D0*rhoq/cv(1,i))
      dv(2,i)=dv(1,i)/(rgas*cv(1,i))
      dv(3,i)=Sqrt(g1cp*dv(2,i))
      dv(4,i)=gamma
      dv(5,i)=cpgas
    enddo
! Navier-Stokes equations
  else
    s1=110D0
    s2=288.16D0
    s12=1D0+s1/s2
```

```
      cppr=cpgas/prlam
      do i=1,nnodes
        rhoq=cv(2,i)*cv(2,i)+cv(3,i)*cv(3,i)
        dv(1,i)=gam1*(cv(4,i)-0.5D0*rhoq/cv(1,i))
        dv(2,i)=dv(1,i)/(rgas*cv(1,i))
        dv(3,i)=Sqrt(g1cp*dv(2,i))
        dv(4,i)=gamma
        dv(5,i)=cpgas
        rat=Sqrt(dv(2,i)/s2)*s12/(1D0+s1/dv(2,i))
        dv(6,i)=refvisc*rat
        dv(7,i)=dv(6,i)*cppr
      enddo
    endif
  end subroutine DependentVarsAll
```

Example 3: Flow over a Simple Airfoil

This next example is also inviscid (i.e., Euler), as described by the input file
(n0012.inp). The core grid and pressure field are shown in the following figure:

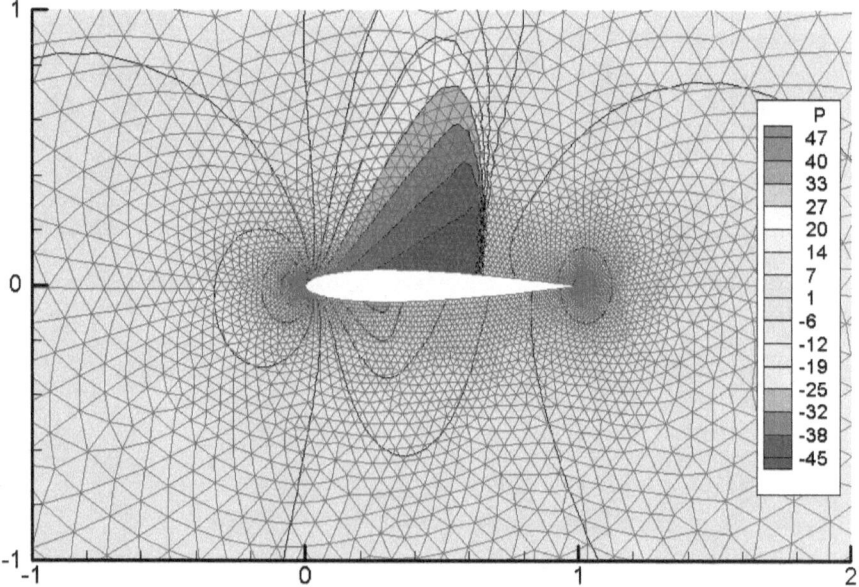

Roe's flux-splitting scheme and average are implemented in the following 5
subroutines and 12 lines of code:

```
! - Roe's flux-difference splitting scheme (upwind)
subroutine DissipRoe1(beta)
subroutine DissipRoe1Prec(beta)
subroutine DissipRoe2(beta)
subroutine DissipRoe2Prec(beta)
subroutine LimiterRefvals
! - Roe's average
```

```
rav=Sqrt(rl*rr)
gam1=0.5D0*(dv(4,i)+dv(4,j))-1D0
dd=rav/rl
dd1=1.D0/(1D0+dd)
uav=(ul+dd*ur)*dd1
vav=(vl+dd*vr)*dd1
pav=(pl+dd*pr)*dd1
tav=(tl+dd*tr)*dd1
hav=(hl+dd*hr)*dd1
q2a=uav*uav+vav*vav
cav=Sqrt(gam1*(hav-0.5D0*q2a))
uv=uav*nVec(1)+vav*nVec(2)
```

Select velocity vectors (1 out of 9) are shown below:

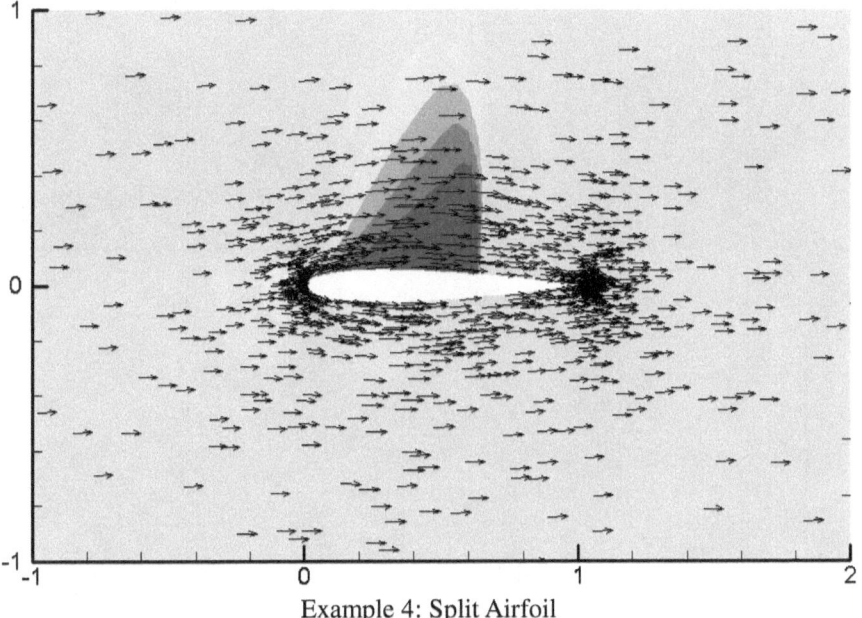

Example 4: Split Airfoil

The fourth example is a split airfoil and also inviscid (see line 25 of n4415.inp). For all five examples, the number of Jacobi iterations for residual smoothing is set to 2 (see line 66). This is implemented in subroutine Irsmoo() and is one of the parameters you can experiment with, should you encounter erratic iterations or an excessive number of iterations. As you read through the input files for these 5 examples, it should become apparent that there are a great many adjustable parameters that might impact the results. While a few of these seem familiar and clear, most are unfamiliar or vague. Some of this ambiguity arises from the fact that this and most every other CFD code is perpetually in the experimental/development stage. Don't expect to find any such code with which you can throw together a model and come up with accurate results.

The pressure field and mesh is shown in this next figure:

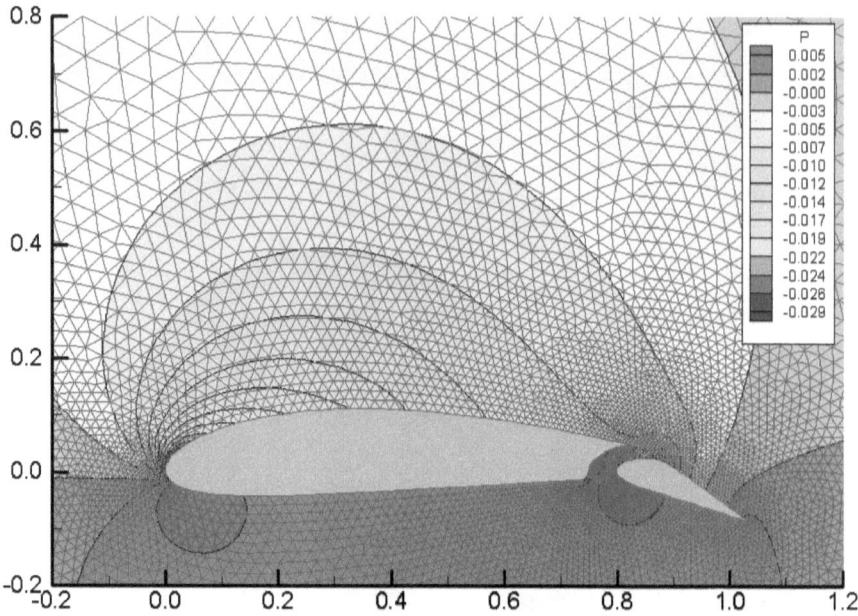

The velocity vectors (1 out of 9) are shown in this next figure:

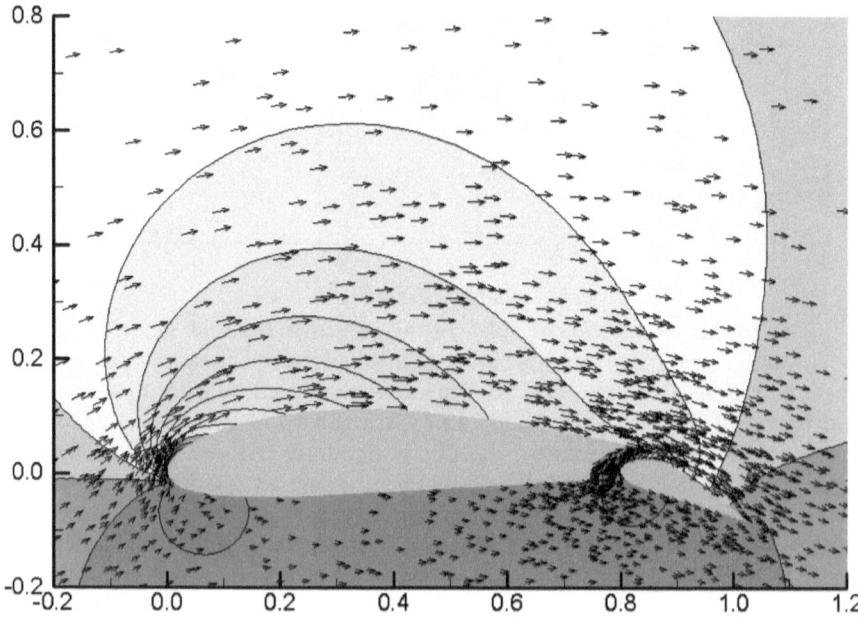

Example 5: Flow over a Turbine Blade

The last example provided with unstruct2d is inviscid flow (Euler) over a single turbine blade. The input (vki1.inp) and mesh (vki1.ugr) files describe the problem. The grid is shown in this first figure:

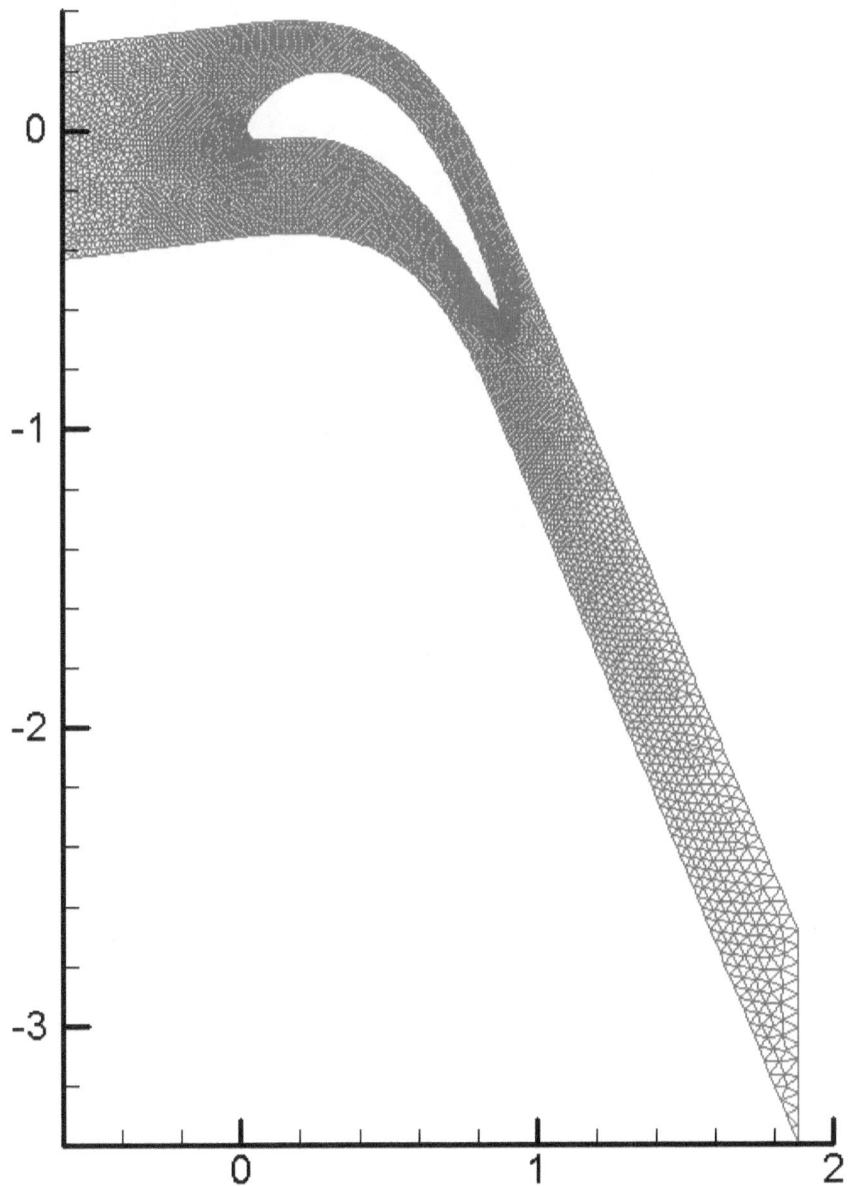

As before, the mesh is refined around the points of interest (i.e., the leading and trailing edges):

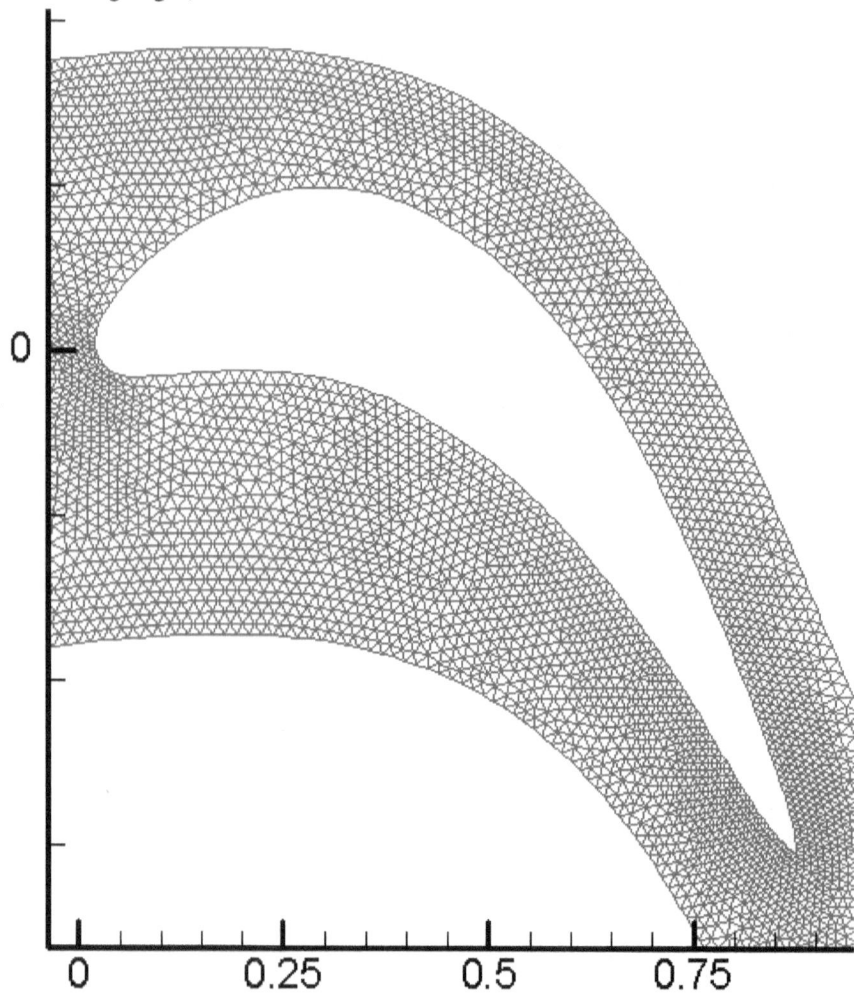

The gradients are calculated in subroutine GradientsVisc():

```
subroutine GradientsVisc
!> Computes gradients of the density, u, v, pressure and
   of the temperature
!! with respect to the x- and y-coordinates. Gradients
   are evaluated at
!! the grid nodes.
   do ie=1,nedint
     i=edge(1,ie)
     j=edge(2,ie)
```

```
! - average of variables
    rav=0.5D0*(cv(1,i)+cv(1,j))
    uav=0.5D0*(cv(2,i)/cv(1,i)+cv(2,j)/cv(1,j))
    vav=0.5D0*(cv(3,i)/cv(1,i)+cv(3,j)/cv(1,j))
    pav=0.5D0*(dv(1,i)+dv(1,j))
    tav=0.5D0*(dv(2,i)+dv(2,j))
! - gradients (divided later by the volume)
    fcx(1)=rav*sij(1,ie)
    fcx(2)=uav*sij(1,ie)
    fcx(3)=vav*sij(1,ie)
    fcx(4)=pav*sij(1,ie)
    fcx(5)=tav*sij(1,ie)
    fcy(1)=rav*sij(2,ie)
    fcy(2)=uav*sij(2,ie)
    fcy(3)=vav*sij(2,ie)
    fcy(4)=pav*sij(2,ie)
    fcy(5)=tav*sij(2,ie)
    gradx(1,i)=gradx(1,i)+fcx(1)
    gradx(2,i)=gradx(2,i)+fcx(2)
    gradx(3,i)=gradx(3,i)+fcx(3)
    gradx(4,i)=gradx(4,i)+fcx(4)
    gradx(5,i)=gradx(5,i)+fcx(5)
    gradx(1,j)=gradx(1,j)-fcx(1)
    gradx(2,j)=gradx(2,j)-fcx(2)
    gradx(3,j)=gradx(3,j)-fcx(3)
    gradx(4,j)=gradx(4,j)-fcx(4)
    gradx(5,j)=gradx(5,j)-fcx(5)
    grady(1,i)=grady(1,i)+fcy(1)
    grady(2,i)=grady(2,i)+fcy(2)
    grady(3,i)=grady(3,i)+fcy(3)
    grady(4,i)=grady(4,i)+fcy(4)
    grady(5,i)=grady(5,i)+fcy(5)
    grady(1,j)=grady(1,j)-fcy(1)
    grady(2,j)=grady(2,j)-fcy(2)
    grady(3,j)=grady(3,j)-fcy(3)
    grady(4,j)=grady(4,j)-fcy(4)
    grady(5,j)=grady(5,j)-fcy(5)
  enddo
! divide by the control volume ------------------------
  ----------------------
  do i=1,nndint
    gradx(1,i)=gradx(1,i)/vol(i)
    gradx(2,i)=gradx(2,i)/vol(i)
    gradx(3,i)=gradx(3,i)/vol(i)
    gradx(4,i)=gradx(4,i)/vol(i)
    gradx(5,i)=gradx(5,i)/vol(i)
    grady(1,i)=grady(1,i)/vol(i)
etc...
  enddo
```

The pressure field and select velocities (1 of 9) are shown in this last figure:

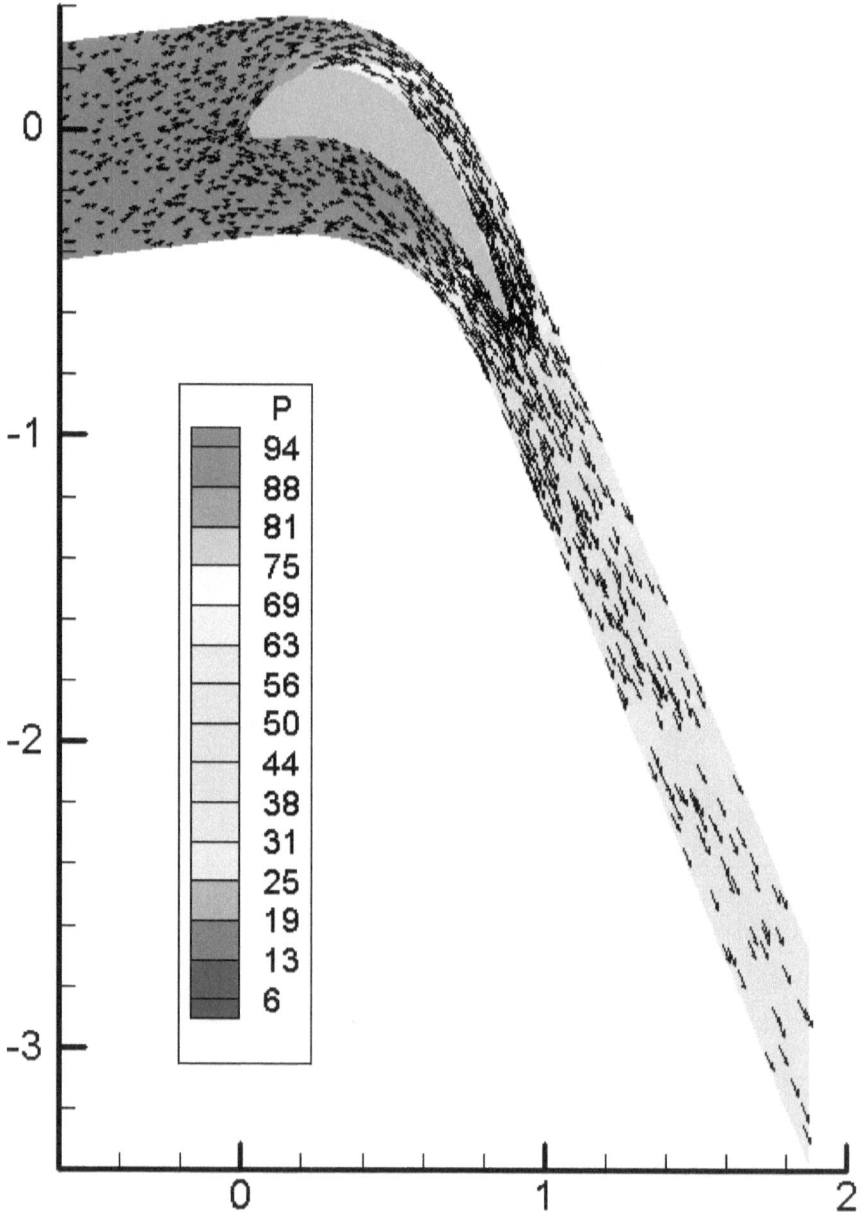

Mohammadi's NSC2KE

Before we leave the FVM, we will consider a similar model that also uses this same triangular mesh. NSC2KE is a Finite-Volume Galerkin program for computing 2D and axisymmetric flows on unstructured meshes written by Bijan Mohammadi. Roe, Osher, and kinetic approaches are available to solve the Euler part of the governing equations. A k-epsilon model is used for turbulence, which is why we cover this program here. We will discuss the topic of turbulence in Chapter 6. A fourth order Runge-Kutta solver is used for time stepping. The original unmodified code can be downloaded from several locations, including GitHub:

https://github.com/cpraveen/nsc2ke

The source code has been reorganized for compatibility and the output files have been modified to target Tecplot® or TP2. You can find all of the associated files, including source, executable, input, output, and documentation in the online archive in folder examples\nsc2ke. The source (nsc2ke.for) is FORTRAN-77. The precompiled executable (nsc2ke.exe) will run on any version of Windows®. There are two input files (data.inp and mesh.inp) for the single example, which is the same problem as previous Example 3, flow over a NACA 0012 airfoil. The grid is shown in the figure below:

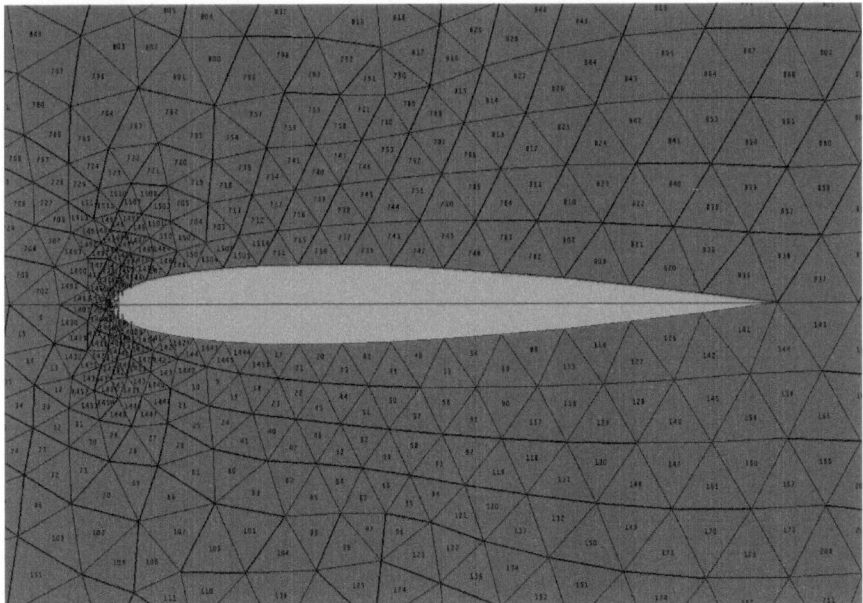

A Tecplot™ layout file (nsc2ke.lay) can also be found in this same folder, which has been adjusted to best display the results. The pressure field is written out at the vertices of the elements and the velocity vectors are written out at the center of the elements.

Results in the vicinity of the airfoil are shown in the following figure:

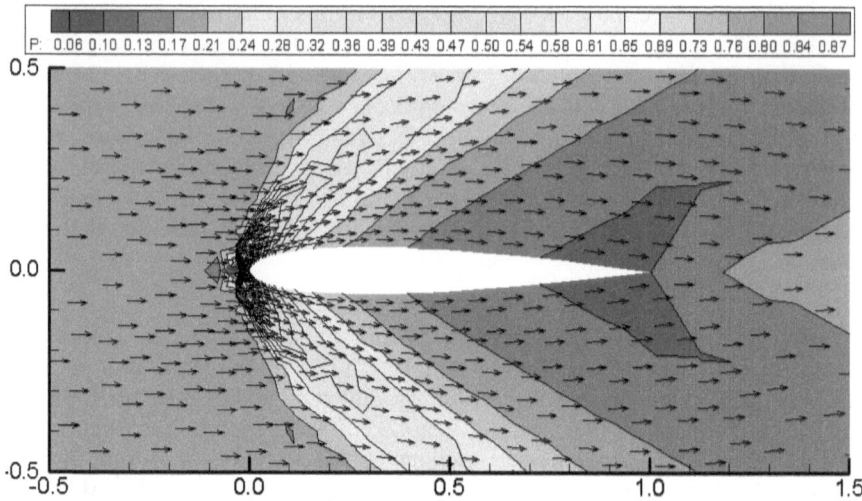

The pressure pattern is typical for near sonic flow. An expanded view showing log(p):

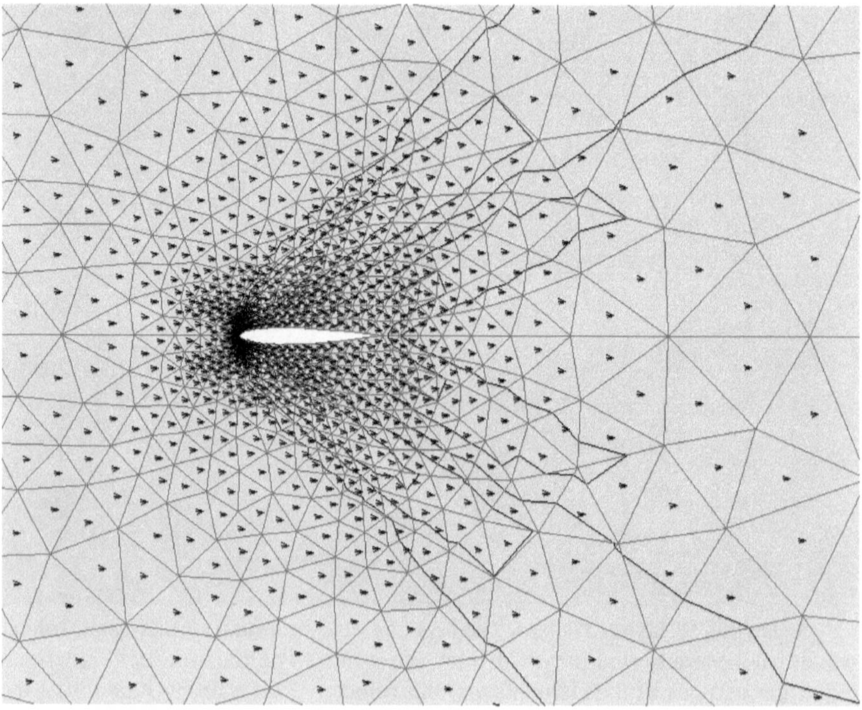

Chapter 5. Finite Element Method

I presume here that you have some understanding of finite element methods. If not, I would suggest you familiarize yourself with the basics before continuing. Before we delve into exactly how it's done, we will consider a simple illustration. Hyung-Chun Lee has provided an F90 code (tcell.f), which will serve this purpose well. You can find this interesting little program at several web sites, including John Burkardt's FORTRAN trove.

http://people.math.sc.edu/Burkardt/

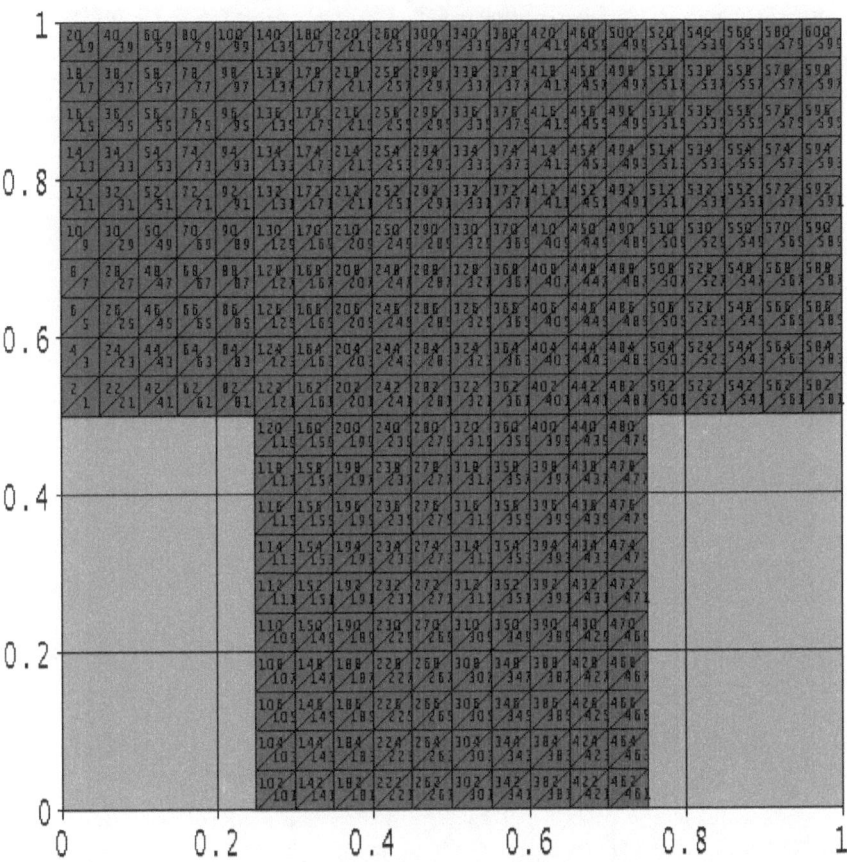

The code is explained in several articles available on the Web by Burkardt, Gunzburger, and Lee.[16] Basically, it first creates a set of triangular elements that

---

[16] Burkardt, J., Gunzburger, M., and Lee, H-C, "POD and CVT-Based Reduced-Order Modeling of Navier–Stokes Flows," Computational Methods in Applied Mechanics and Engineering, Vol. 196, pp. 337–355, 2006.

fill a T-shape and then solves the modified Navier-Stokes equations for pressure and velocity. I have completely rewritten the code in C so that it dynamically allocates the arrays, doesn't require a FORTRAN compiler, and creates output files ready for TP2 or Tecplot™. You will find the original annotated code plus all of the associated files in the online archive in folder examples\tcell. The elements are shown in the preceding figure and the velocity vectors in the following:

results of tcell.c with BC=1

You can change several parameters in the code and recompile to produce variants. The flow pattern above (BC=1) has the flow entering on the top left and exiting on the top right. BC=2 shown in this next figure has the flow entering at the bottom and leaving on the top right. It also shows the pressure contours. As before, several output files are created, including a data and layout for Tecplot™.

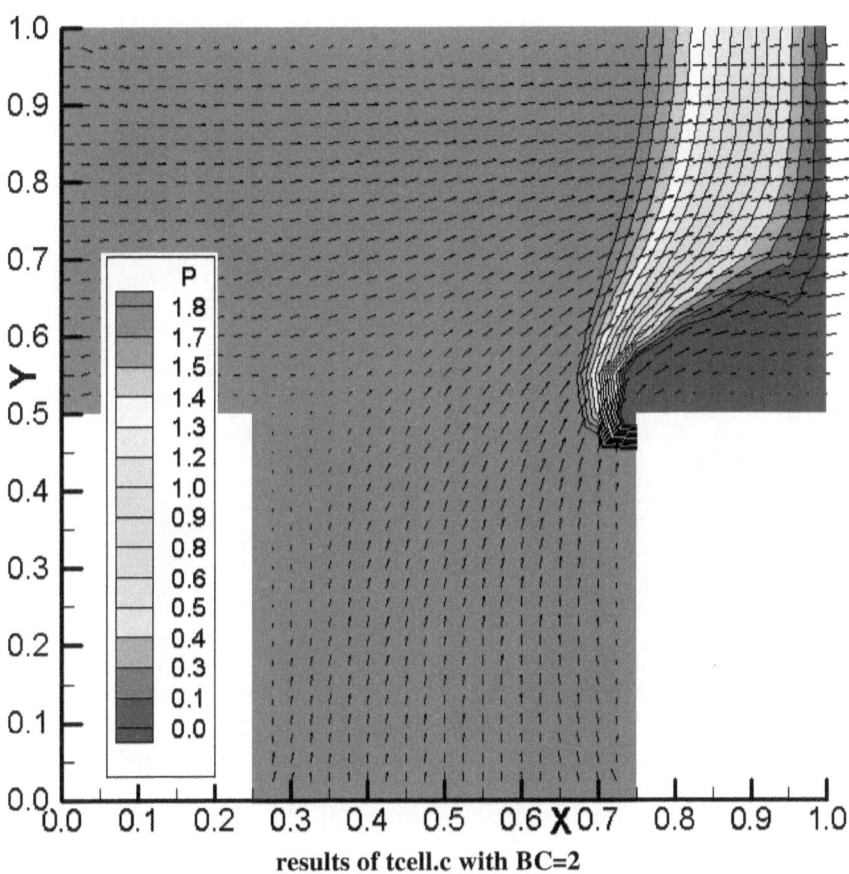

results of tcell.c with BC=2

Of course, we could solve this simple geometry with the FDM code from Chapter 3 by making a few adjustments. You will also find this modified code (fdm.c) in the same folder with tcell.c. The results are similar.

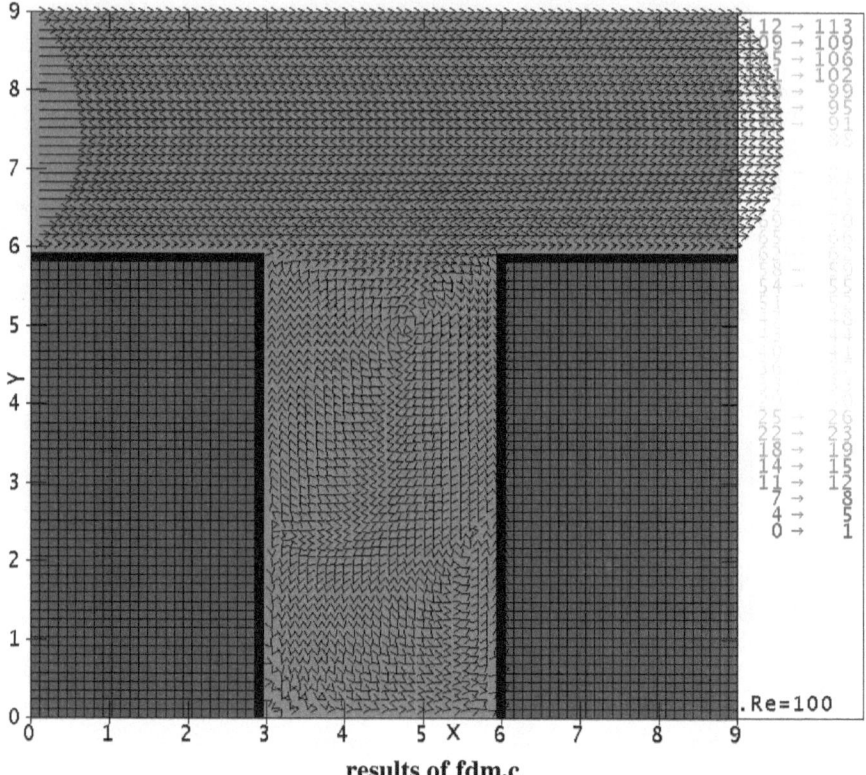

results of fdm.c

At Burkardt's web site you can also get a similar code (hcell.f) that creates an H-shaped region. You will find those files (hcell.f, hcell.for, and hcell.c) plus the associated plot files in the same folder. The FDM code can be modified for the differing pattern.

I have adjusted the dimensions and also the boundary conditions (hcell.c) to produce a more interesting pattern than the original code.

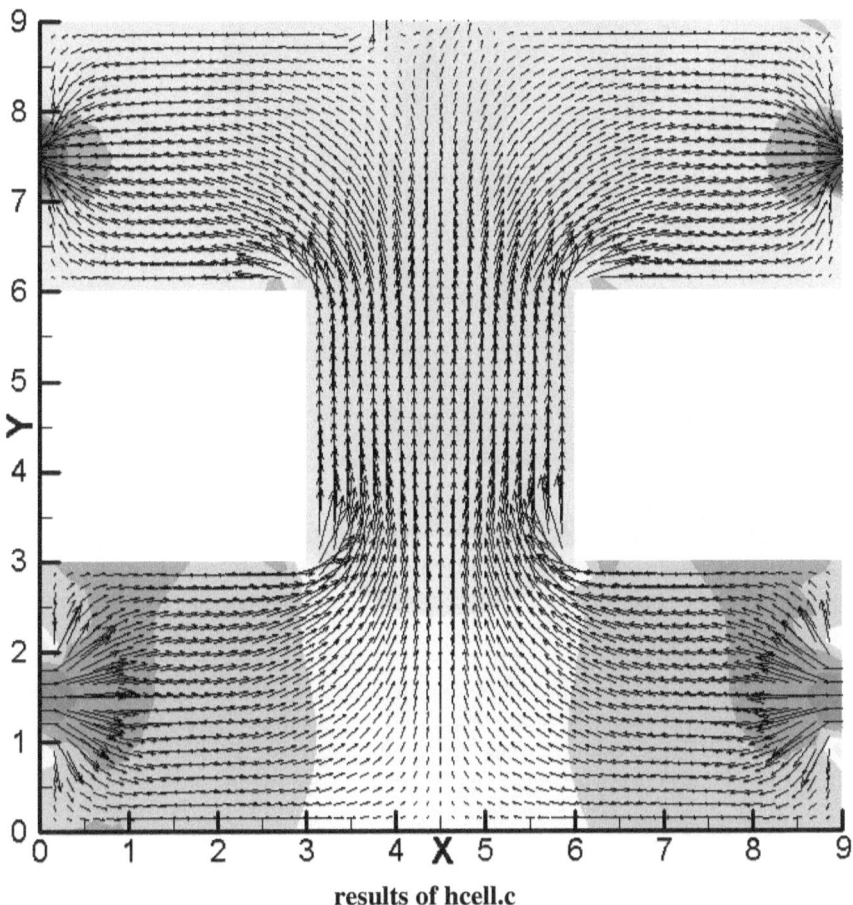

results of hcell.c

Taylor-Hood Elements

Now let's take a closer look at what Lee is doing. First, the triangular elements are called Taylor-Hood:

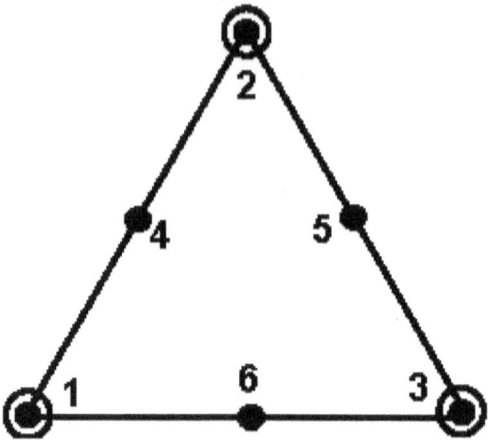

We use two basis functions: 1) linear and 2) quadratic. The linear basis function has three terms $(a+b*x+c*y)$ and utilizes the three corner nodes. The quadratic basis function has six terms $(a+b*x+c*y+d*x^2+e*x*y+f*y^2)$ and utilizes all six nodes. It is fortuitous that 3 and 6 correspond to the binomial terms in the simplest two-variable (i.e., x, y) expansion. As noted in the preceding chapter, the partial differential equations we hope to solve have both first and second order terms. With the Taylor-Hood method, we use the linear basis function for the first order terms and the quadratic basis function for the second order terms. In the streamlined code (tcell.c), lbf() is the linear basis function and qbf() is the quadratic basis function.

```
double lbf(double x,double y,int iq)
  { /* linear basis function */
  if(iq==1)
    return(1.-x-y);
  if(iq==2)
    return(x);
  if(iq==3)
    return(y);
  return(0.);
  }
void qbf(double x,double y,int i,double*bb, double*bx,
    double*by)
  { /* quadradic basis function */
  if(i==1)
    {
    *bb=(1.-x-y)*(1.-2.*x-2.*y);
    *bx=-3.+4.*x+4.*y;
```

```
     *by=-3.+4.*x+4.*y;
     etc.
}
```

The equations are put into matrices, a[]*f[]=g[], where a[] is the left-hand side, g[] is the right-hand side, and f[] is the solution (i.e., pressure and the two velocity components). The position of each solution component (p, u, or v) is kept in an array (indx[] for u and v or insc[] for p). The results are accessed as illustrated below, where they're written to the Tecplot™ data file:

```
for(ic=0;ic<np;ic++)
{
u=f[indx[2*ic]];
v=f[indx[2*ic+1]];
p=f[insc[ic]];
fprintf(fp,"%1G %1G %1G %1G
%1G\n",xc[ic],yc[ic],p,u,v);
}
```

The matrix a[] is *banded*, that is, it's not a full **nxn** matrix; rather, it's mostly zeroes. The span is determined by the number of neighbors for each node, that is, the other nodes that appear in at least one element with a particular node. The positioning of the elements and ordering of indices facilitate solving the resulting banded matrix for the desired values (p, u, and v at each node). The matrix is solved by SolveBanded(), which calls Eliminate() repeatedly for the rows. These functions utilize several vector (i.e., array) operation functions from LINPACK, including dot product, scale, axpy (a times x plus y), and locate index of largest value. Relative runtimes are listed in the following table:

function	time	cumulative	description
axpy	66.3985%	66.3985%	vector op: y[]+a*x[]
Eliminate	25.7623%	92.1609%	row elimination
NavierStokes	6.5383%	98.6992%	Navier-Stokes
qbf	0.5083%	99.2075%	quadratic basis function
lbf	0.4760%	99.6834%	linear basis function
scale	0.0998%	99.7832%	vector op: v[]*a
SolveBanded	0.0936%	99.8768%	banded matrix solver
allocate	0.0439%	99.9207%	memory allocation
ivmax	0.0318%	99.9525%	vector op: index of largest element
bc	0.0257%	99.9782%	boundary condition
trans	0.0097%	99.9879%	variable transform
WriteV2D	0.0058%	99.9937%	file velocity vectors
Write2DV	0.0041%	99.9978%	file elements
CreateGrid	0.0019%	99.9997%	create grid
main	0.0003%	100.0000%	main

We see that 66.3985% of the time is consumed in axpy—and that's after I optimized it, eliminating the optional non-unity ($\neq$1) increment. The row elimination process (including axpy) consumes 92.1609% of the time. The Navier-Stokes equations, per se, plus the linear and quadratic basis functions account for a trivial fraction of the total time. This is typical for FEM codes. The same analysis for the FDM codes in Chapter 3 would reveal that most of the time is consumed calculating the differences over and over again. For more details, see *profiling* at the end of Appendix B.

FEM Rationale

Fundamental concepts behind the Finite Element Method (FEM) are often lost in the details of whose approach is used and how. The Finite Difference Method (FDM) is much more intuitive. We approximate each differential with differences, combine these to form the governing equations, and crunch through the calculations in order to obtain a result. We can't follow this same approach with FEM for several reasons.

First, when we consider a single element, there's no direct path to calculating the differentials. Second, even if we are clever enough to come up with formulas for the differentials, it's not at all clear where these might apply: in the middle of the element, over the entire element, at one of the corners, along one of the edges, etc. Trudging through this process, as with the FDM, we would end up with a mismatch between the number of equations and unknowns.

With the FDM, we can build an algebraic equation for each unknown at each cell. With the FEM, we don't have enough information with a single element to calculate anything. With the FEM, we must assemble information from all of the elements before we have enough information to solve for any one of them. This is why at the end of the preceding chapter we compared the function times between the FDM and FEM. Solving for a flow field with the FDM mostly involves calculating differences, while the FEM will mostly involve solving matrices. In short, FDM calculations focus on small groups of a few nodes, while FEM calculations focus on the collective group of all nodes (i.e., the ensemble).

With the FDM, we want each differential to accurately represent a part of the governing equation(s). If these do, we consider the governing equation to be accurately satisfied within each element. With the FEM, we approach this same goal from a different direction. The governing equation(s) typically equals zero or we can rearrange it so that this is the case. If we integrate our as-yet-to-be-defined representation of the governing equation(s) over an element (2D area or 3D volume), we obtain a measure of how well we're achieving this goal.

If we combine these integrals for all elements, we get the total discrepancy. We also get an expression for our goal that includes all of the unknowns. The total discrepancy will be minimized when the partial derivatives with respect to each unknown reaches zero. These partial derivatives inherently yield the same number of equations as unknowns. The question then becomes, "How do we accomplish this?"

Basis Functions

If we're going to integrate the governing equation over each element, we need some representation of the parameters, which vary over an element in a non-trivial way. For flow with constant properties, this is would be the pressure and velocity components. We may add more parameters later in order to accurately handle such things as turbulence. This is where the linear and quadratic basis functions from the preceding chapter come into the process. A search of the Web will reveal that a variety of basis functions have been used with the FEM, but these are the simplest, which is why they were presented first.

Integration

We could analytically integrate some relationships, though more complicated ones might be easier to integrate numerically. Various methods of integration are covered in my book, *Numerical Calculus*, so we will not dwell on the details in this text. The one topic we will discuss is Green's Lemma. This very useful principle transforms the integral over an area to one around the boundary. It's a whole lot easier to integrate around the sides of a triangle than over the area. Green's Lemma, along with a validation example code, can be found in Appendix E.

Equation Building

We now consider a snippet from tcell.c:

```
1 for(i=1;i<=6;i++)
2   {
3   qbf(x,y,i,&bb,&tbx,&tby);
4   bx=tbx*xix+tby*etax;
5   by=tbx*xiy+tby*etay;
6   bbl=lbf(x,y,i);
7   for(j=1;j<=3;j++)
8     {
9     f[ii]+=csim*((un[0]*unx[0]+un[1]*uny[0])*bb)
            *ar+uqp*bb*ar/deltat;
10    for(k=1;k<=6;k++)
11      {
12      qbf(x,y,k,&bbb,&tbbx,&tbby);
13      bbx=tbbx*xix+tbby*etax;
14      bby=tbbx*xiy+tbby*etay;
15      bbbl=lbf(x,y,k);
16      a[ii][jj]=visc*(by*bby+bx*bbx)+(bbb*unx[0]*bb)
              *csim+bb*bbx*un[0]+bb*bby*un[1]+bb*bbb/deltat;
```

The linear basis function, lbf(), is called on line 6, returning a single value. The quadratic basis function, qbf(), is called on lines 3 and 12, returning three values (viz., bb, tbx, and tby). The necessary analytical integration and differentiation steps are built into the respective functions, so that they return the appropriate terms. The first for (i) loops over the 6 nodes. The second for (j) loops over the 3 equations (p, u, and v), accumulating right-hand side terms

61

(f[ii]+=…). The third for (k) loops over the 6 nodes again, providing left-hand side terms (a[ii][jj]=…), representing the contribution of each node to each equation. We see the time step (deltat) in lines 9 and 16 and also the kinematic viscosity (visc) in line 16, which reflect the governing equations (i.e., continuity and Navier-Stokes, that is, conservation of mass and linear momentum).

I have changed some of the indices in the preceding code to simplify, as these details are not essential to the present discussion. There is considerable bookkeeping required to assemble, address, and solve the a[] and f[] matrices in an efficient manner, using as little memory as possible. A discussion of elements and bookkeeping can be found in Appendix F. We will consider index building and management after delving further into the integration process.

Consider the following integral, which follows from Green's Lemma:

$$\iint x^a y^b \, dA = \oint \frac{x^a y^{b+1}}{b+1} \, dS \qquad (5.1)$$

We could have just as easily integrated with respect to x. For a triangular element, this becomes:

$$\iint x^a y^b \, dA = \sum_{i=1}^{3} \int_{i}^{i+1} \frac{x^a y^{b+1}}{b+1} \, dS \qquad (5.2)$$

The coordinates (x,y) vary linearly from one point to the next:

$$x = x_1(1-\sigma) + x_2\sigma$$
$$y = y_1(1-\sigma) + y_2\sigma \qquad (5.3)$$

The relationship between σ and dS is given by the following equation, where σ varies from 0 to 1 along each side.

$$d\sigma = \frac{dS}{\sqrt{(x_2-x_1)^2 + (y_2-y_1)^2}} \qquad (5.4)$$

The first segment (point 1 to point 2) expands to:

$$\int_0^1 [x_1(1-\sigma) + x_2\sigma]^a \left[\frac{y_1(1-\sigma) + y_2\sigma}{b+1} \right]^{b+1} \delta \, d\sigma \qquad (5.5)$$

where δ is the denominator of 5.4:

$$\delta = \sqrt{(x_2-x_1)^2 + (y_2-y_1)^2} \qquad (5.6)$$

Equation 5.5 has no closed-form solution for arbitrary values of a and b. For $a=b=0$, the integral in 5.5 becomes the standard formula for the area of a triangle:

62

$$\iint dA = A = \frac{(x_3 - x_2)(y_2 - y_1) - (x_2 - x_1)(y_3 - y_2)}{2} \qquad (5.7)$$

For *a=1, b=0*, the integral becomes:

$$\iint x dA = \frac{(x_1 + x_2 + x_3)}{6}[(y_3 - y_2)x_1 + (y_1 - y_3)x_2 + (y_2 - y_1)x_3] \quad (5.8)$$

For *a=0, b=1*, the integral becomes:

$$\iint y dA = \frac{(y_1 + y_2 + y_3)}{6}[(x_2 - x_3)y_1 + (x_3 - x_1)y_2 + (x_1 - x_2)y_3] \quad (5.9)$$

It should now be a little more clear how the formulas in the snippets of the linear basis function lbf() and quadratic basis function qbf() in tcell.c, listed in the preceding chapter, arise. We form the basis functions using the unknown values of p, u, and v at each of the 3 or 6 points in each element. We substitute these simple linear or quadratic formulas into the governing equations and then integrate around the three sides of the element, from point 1 to 2 to 3 and back to point 1. We take the derivative with respect to p1, p2, p3, u1, u2, u3, u4, u5, u6 v1, v2, v3, v4, v5, and v6, then put these into the left-hand side rectangular banded matrix. Constants go in the right-hand side columnar matrix.

Simplified/Modified Variants

If any of the terms aren't linear in the unknowns (i.e., we have terms like p1p2, p1u1, u2v3, etc.), we use the current values to make the terms linear. If this is the case, we will have to update the left and right-hand sides periodically in order to obtain a solution. This will no doubt significantly increase the run times. This also gives rise to the many variants called *simplified* or *modified*. There are countless ways to sidestep the nonlinear term problem, but these will necessarily result in a different solution.

In Lee's T-cell program, lbf() and qbf() are called from the function NavierStokes(), which builds the equations (i.e., matrices a[] and f[]).[17] The main calling loop is listed below:

```
t=0.;
do{
   memcpy(g,f,neqn1*sizeof(double));
   memset(f,0,neqn1*sizeof(double));
   NavierStokes();
   t+=deltat;
   }
}while(t<tend);
```

---

[17] Note: In rewriting the code, I have also renamed the functions to be more intuitive. These may be found in tcell.c. The functions were originally called refbsp(), refbqf(), and nstoke(), respectively, in tcell.f.

Here, g[] contains the previous values of f[], memcpy is a hardware-level fast block copy, and memset is a hardware-level fast fill. Double-precision (64-bit floating-point) 0.0 just happens to be all zeroes in binary; otherwise, we couldn't use this function to set a block of floating-point values. For instance, this doesn't work for 1.0. The function NavierStokes() is inside the time loop, so matrices a[] and f[] are rebuilt at each step, using the current (old) values, which is why we must save them in g[].[18]

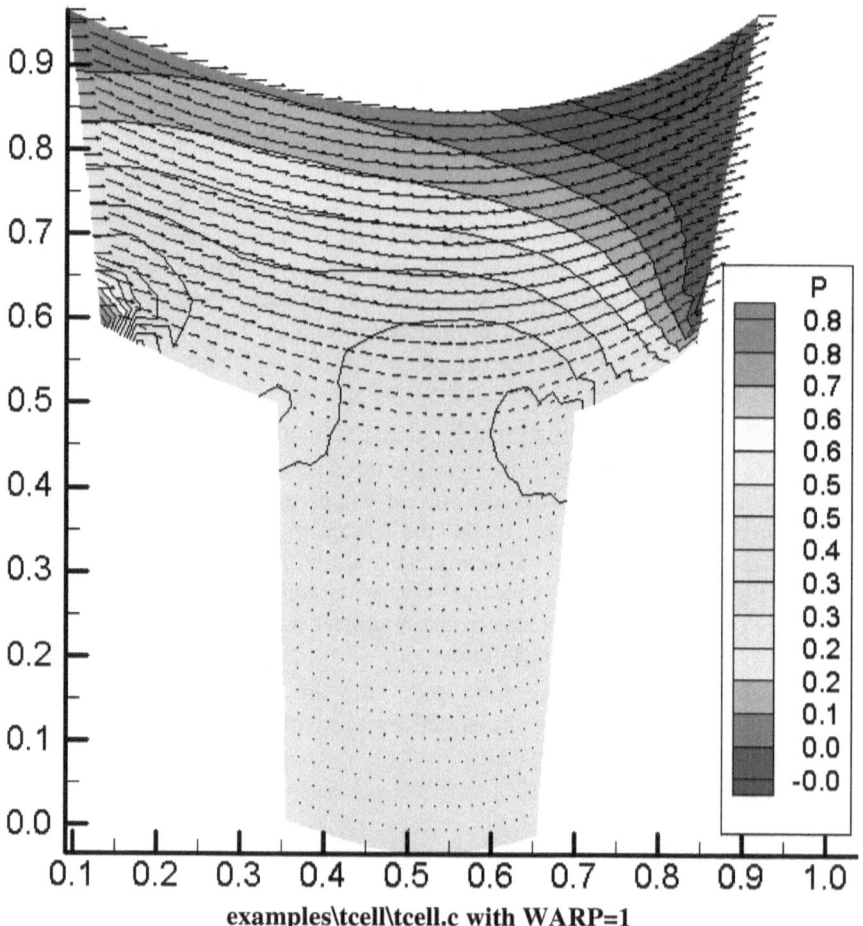

examples\tcell\tcell.c with WARP=1

Besides fitting complex boundaries, another advantage of the FEM (triangular or tetrahedral elements) over the FDM is that the governing equations naturally stretch but remain valid as the nodes shift. We can warp the

---

[18] The columnar matrix, g[], in the modernized code (tcell.c) is called uold[] in the original FORTRAN (tcell.f).

domain and still solve the problem without having to adjust the finite differences to compensate.

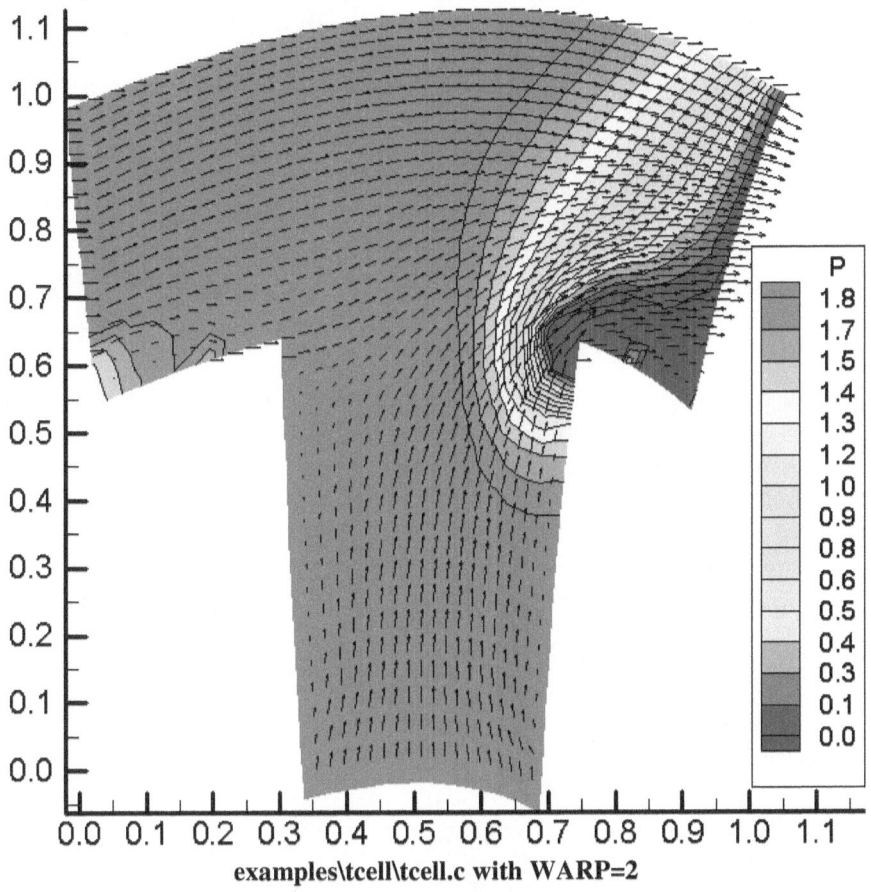

examples\tcell\tcell.c with WARP=2

<u>Approximated Governing Equations</u>

We can now use the basis functions to write the approximated governing equations for a single element. The continuity equation (3.4) becomes:

$$
\begin{aligned}
\frac{\partial p}{\partial t} = & -(u1 + u2\,x + u3\,y + u4\,x^2 + u5\,x\,y + u6\,y^2)\,p2 \\
& -(v1 + v2\,x + v3\,y + v4\,x^2 + v5\,x\,y + v6\,y^2)\,p3 \\
& -(p1 + p2\,x + p3\,y)\,(u2 + 2\,u4\,x + u5\,y + v3 + v5\,x + 2\,v6\,y)
\end{aligned} \tag{5.10}
$$

The X-momentum equation (3.2x) becomes:

$$\frac{\partial u}{\partial t} = -(u1 + u2\,x + u3\,y + u4\,x^2 + u5\,x\,y + u6\,y^2)\,(u2 + 2\,u4\,x + u5\,y)$$
$$- (v1 + v2\,x + v3\,y + v4\,x^2 + v5\,x\,y + v6\,y^2)\,(u3 + u5\,x + 2\,u6\,y)$$
$$- \frac{p2}{\rho} + \frac{(2\,u4 + 2\,u6)\,\mu}{\rho} \qquad (5.11)$$

The Y-momentum equation (3.2y) becomes:

$$\frac{\partial v}{\partial t} = -(u1 + u2\,x + u3\,y + u4\,x^2 + u5\,x\,y + u6\,y^2)\,(v2 + 2\,v4\,x + v5\,y)$$
$$- (v1 + v2\,x + v3\,y + v4\,x^2 + v5\,x\,y + v6\,y^2)\,(v3 + v5\,x + 2\,v6\,y)$$
$$- \frac{p3}{\rho} + \frac{(2\,v4 + 2\,v6)\,\mu}{\rho} \qquad (5.12)$$

We now integrate these over an element (i.e., following Equations 5.2 through 5.9), which yields an expression for the conservation of mass and momentum for a single element:

```
double C(double x1,double x2,double x3,double y1,
    double y2,double y3,double p1,double p2,double p3,
    double u1,double u2,double u3,double u4,double u5,
    double u6,double v1,double v2,double v3,double v4,
    double v5,double v6)
    {
    return((y3*x2-y1*x2+y1*x3+y2*x1-y2*x3-y3*x1)*
    (3.*p2*u4*x3*x3+p3*u5*y3*y3+p2*u6*y1*y1
    +6.*p2*u1+4.*p1*v6*y1+p3*v4*x3*x3+p3*u5*y2*y2
    +p3*v5*y1*x2+2.*p3*v5*y1*x1+2.*p3*v5*y2*x2
    +p3*v5*y2*x1+p3*v4*x2*x1+3.*p3*v6*y1*y2
    +p3*u5*y1*y2+p2*u5*y1*x2+2.*p2*u5*y1*x1
    +2.*p2*u5*y2*x2+p2*u5*y2*x1+p2*v6*y1*x2
    +2.*p2*v6*y1*x1+2.*p2*v6*y2*x2+p2*v6*y2*x1
    +p3*u4*y1*x2+2.*p3*u4*y1*x1+2.*p3*u4*y2*x2
    +p3*u4*y2*x1+3.*p2*u4*x2*x1+p2*u6*y1*y2
    +p2*v5*x2*x1+p3*v4*x2*x2+p3*v4*x1*x1
    +3.*p3*v6*y1*y1+3.*p3*v6*y2*y2+p3*u5*y1*y1
    +3.*p2*u4*x2*x2+3.*p2*u4*x1*x1+p2*u6*y2*y2
    +p2*v5*x2*x2+p2*v5*x1*x1+2.*p1*u5*y1
    +2.*p1*u5*y2+4.*p1*v6*y2+2.*p3*u2*y1
    +2.*p3*u2*y2+4.*p3*v3*y1+4.*p3*v3*y2
    +3.*p3*v6*y3*y3+p3*v5*y2*x3+2.*p3*v5*y3*x3
    +p3*v5*y3*x2+p3*v4*x3*x2+3.*p3*v6*y2*y3
    +p3*u5*y2*y3+p2*u5*y2*x3+2.*p2*u5*y3*x3
    +p2*u5*y3*x2+p2*v6*y2*x3+2.*p2*v6*y3*x3
    +p2*v6*y3*x2+p3*u4*y2*x3+2.*p3*u4*y3*x3
    +p3*u4*y3*x2+3.*p2*u4*x3*x2+p2*u6*y2*y3
    +p2*v5*x3*x2+p2*u6*y3*y3+p2*v5*x3*x3
```

```
       +2.*p1*u5*y3+4.*p1*v6*y3+2.*p3*u2*y3
       +4.*p3*v3*y3+p3*v5*y3*x1+p3*v5*y1*x3
       +p3*v4*x1*x3+3.*p3*v6*y3*y1+p3*u5*y3*y1
       +p2*u5*y3*x1+p2*u5*y1*x3+p2*v6*y3*x1
       +p2*v6*y1*x3+p3*u4*y3*x1+p3*u4*y1*x3
       +3.*p2*u4*x1*x3+p2*u6*y3*y1+p2*v5*x1*x3
       +4.*x3*p2*u2+2.*p2*v3*x2+4.*p2*u2*x2
       +2.*p1*v5*x2+4.*p1*u4*x2+2.*p2*u3*y3
       +2.*p3*v2*x2+2.*x1*p2*v3+2.*x1*p3*v2
       +4.*x1*p1*u4+2.*x1*p1*v5+4.*x1*p2*u2
       +2.*y2*p2*u3+2.*y1*p2*u3+4.*x3*p1*u4
       +2.*x3*p2*v3+2.*x3*p1*v5+2.*x3*v2*p3
       +6.*v3*p1+6.*u2*p1+6.*v1*p3)/12.);
  }
```

Then we take the partial derivatives with respect to p1, p2, p3, u1, u2, u3, etc. As some of these are quite lengthy, we will only list a select few here. The partial of Continuity (i.e., integrated Equation 5.10) with respect to p1 is:

$$\frac{\partial C}{\partial p1} = \frac{1}{12}(y3\,x2 - y1\,x2 + y1\,x3 + y2\,x1 - y2\,x3 - y3\,x1)\,(2\,u5\,y1$$
$$+ 4\,v6\,y2 + 2\,u5\,y3 + 6\,v3 + 6\,u2 + 4\,v6\,y3 + 2\,x1\,v5 + 4\,x1\,u4$$
$$+ 2\,v5\,x2 + 2\,u5\,y2 + 2\,x3\,v5 + 4\,x3\,u4 + 4\,u4\,x2 + 4\,v6\,y1)$$

(5.13)

The partial of integrated Equation 5.10 with respect to p2 is:

$$\frac{\partial C}{\partial p2} = \frac{1}{12}(y3\,x2 - y1\,x2 + y1\,x3 + y2\,x1 - y2\,x3 - y3\,x1)\,(3\,u4\,x2^2$$
$$+ 4\,x1\,u2 + 4\,x3\,u2 + 3\,u4\,x3^2 + 4\,u2\,x2 + 2\,x3\,v3 + 2\,u3\,y3$$
$$+ 2\,y1\,u3 + 2\,v3\,x2 + 2\,x1\,v3 + v5\,x1^2 + 6\,u1 + u6\,y3^2 + u6\,y2^2$$
$$+ v5\,x3^2 + 3\,u4\,x1^2 + 2\,y2\,u3 + v5\,x2^2 + u6\,y1^2 + u5\,y1\,x2$$
$$+ u5\,y2\,x1 + v6\,y1\,x2 + v6\,y2\,x1 + u6\,y1\,y2 + v5\,x2\,x1 + u5\,y2\,x3$$
$$+ u5\,y3\,x2 + v6\,y2\,x3 + v6\,y3\,x2 + u6\,y2\,y3 + v5\,x3\,x2 + u5\,y3\,x1$$
$$+ u5\,y1\,x3 + v6\,y3\,x1 + v6\,y1\,x3 + u6\,y3\,y1 + v5\,x1\,x3 + 2\,u5\,y1\,x1$$
$$+ 2\,u5\,y2\,x2 + 2\,v6\,y1\,x1 + 2\,v6\,y2\,x2 + 3\,u4\,x2\,x1 + 2\,u5\,y3\,x3$$
$$+ 2\,v6\,y3\,x3 + 3\,u4\,x3\,x2 + 3\,u4\,x1\,x3)$$

(5.14)

To be sure, these equations and their derivation are formidable. I would hate to have to do this without the help of Maple™. The chance of getting any one of them right by hand is slim, let alone all 15. You can find a program (int2d.c) in folder examples\lemma that contains these equations along with a 2D (area) numerical integration, two 1D applications of Green's Lemma (boundary integrals—one on x and a second on y), plus the analytical integration. Random nodes and coefficients are generated and then all four methods are compared to illustrate that we get the same answer. The results are divided by the area to keep the numbers in the same ballpark:

```
testing conservation of mass
  GQ2D LemmaX LemmaY Analyt
   3.4    3.7    3.7    3.7
 139.3  139.3  139.3  139.3
  67.1   67.2   67.2   67.2
 229.8  234.9  234.9  234.9
  27.1   26.9   26.9   26.9
   1.5    4.7    4.7    4.7
  19.9   20.3   20.3   20.3
 124.1  123.4  123.4  123.4
-165.8 -164.6 -164.6 -164.6
  55.9   55.8   55.8   55.8
```

There are 3 principal integrals for each element corresponding to the conservation of mass and linear momentum in two directions (C, X, and Y). There are 15 unknowns for each element (p1 through v6). We build the global matrix by taking the partial of C with respect to p1 through p3 plus the partial of X with respect to u1 through u6 plus the partial of Y with respect to v1 through v6. These are the dominant terms, which will yield the optimal matrix for iterative solution. Setting all of the partial derivatives to zero would yield an indeterminate problem were it not for boundary conditions, which are non-zero. The terms can be optimized to reduce calculation time and eventually end up very much like the previous codes. Below is an excerpt of the code:

```
node[n1-1].ub+=cu1*cb+xu1*xb+yu1*yb;
node[n2-1].ub+=cu2*cb+xu2*xb+yu2*yb;
node[n3-1].ub+=cu3*cb+xu3*xb+yu3*yb;
uu[MA*(n1-1)+ii-1]+=cu1*cu1+xu1*xu1+yu1*yu1;
uu[MA*(n1-1)+ij-1]+=cu1*cu2+xu1*xu2+yu1*yu2;
uu[MA*(n1-1)+ik-1]+=cu1*cu3+xu1*xu3+yu1*yu3;
uv[MA*(n1-1)+ii-1]+=cu1*cv1+xu1*xv1+yu1*yv1;
uv[MA*(n1-1)+ij-1]+=cu1*cv2+xu1*xv2+yu1*yv2;
uv[MA*(n1-1)+ik-1]+=cu1*cv3+xu1*xv3+yu1*yv3;
up[MA*(n1-1)+ii-1]+=cu1*cp1+xu1*xp1+yu1*yp1;
up[MA*(n1-1)+ij-1]+=cu1*cp2+xu1*xp2+yu1*yp2;
up[MA*(n1-1)+ik-1]+=cu1*cp3+xu1*xp3+yu1*yp3;
uu[MA*(n2-1)+ji-1]+=cu2*cu1+xu2*xu1+yu2*yu1;
uu[MA*(n2-1)+jj-1]+=cu2*cu2+xu2*xu2+yu2*yu2;
uu[MA*(n2-1)+jk-1]+=cu2*cu3+xu2*xu3+yu2*yu3;
uv[MA*(n2-1)+ji-1]+=cu2*cv1+xu2*xv1+yu2*yv1;
uv[MA*(n2-1)+jj-1]+=cu2*cv2+xu2*xv2+yu2*yv2;
uv[MA*(n2-1)+jk-1]+=cu2*cv3+xu2*xv3+yu2*yv3;
up[MA*(n2-1)+ji-1]+=cu2*cp1+xu2*xp1+yu2*yp1;
node[n1-1].vb+=cv1*cb+xv1*xb+yv1*yb;
node[n2-1].vb+=cv2*cb+xv2*xb+yv2*yb;
node[n3-1].vb+=cv3*cb+xv3*xb+yv3*yb;
```

Burkardt's FEM2D

For our first implementation of the FEM, we turn again to John Burkardt. You can find the code (fem2d.f90), Windows® executable (fem2d.exe), reference (fem2d.pdf), and three problems in the online archive in folder examples\fem2d. In this folder you will also find a batch file to compile the program and 3 batch files to run each of the examples. The program has been modified to write out one file that can be read by Tecplot™ (fem2d.plt) and two files that can be read by TP2 (fem2d.2dv and fem2d.v2d). I have also prepared a layout file for each of the examples.

The code contains adequate documentation, so it will not all be repeated here. The elements are illustrated and described in subroutine getdu4(). The pressure basis function is evaluated in subroutine bsp(). The flow is solved by subroutine flosol(). This subroutine calls the others necessary to build and solve the matrices. Gauss quadrature is used to integrate around each element. Rather than just initializing values, crunching, and updating again and again to solve the nonlinear equations, a variant of Newton's method is used (see subroutine newton).

FEM2D is not a generic program, rather it is designed to create a specific type of grid and then solve the governing equations. All of the examples are some sort of flow over a bump or step. Based on comments within the code and examples, this program is still under development. The first example (problem6) is flow over a bump. The mesh, pressure field, and velocity vectors are shown in this first figure:

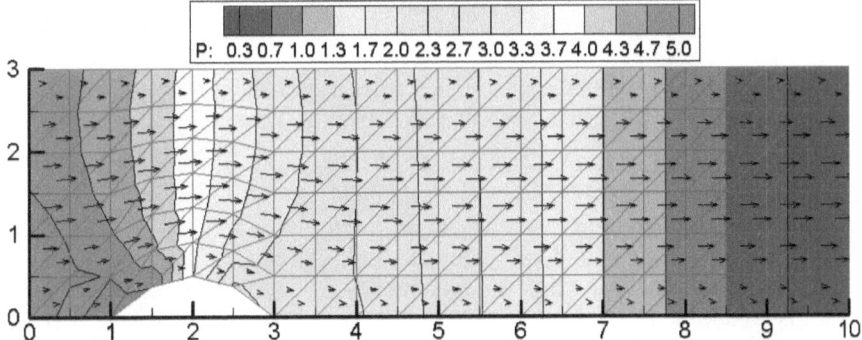

The program has 3 solution options, named: qsolve(), osolve(), and rsolve(). The first two of these call subroutines flosol() and solcon(), while the last calls only flosol(). The first, flosol(), sets up new parameters for Newton iteration and the second, solcon(), utilizes the previous solution. Derivatives of the linear basis function are calculated in subroutine pldx(), while derivatives of the quadratic basis function are calculated in subroutine pqdx(). In that respect, this implementation is similar to what we have seen before.

The next example (problem751) is a slightly different bump and a somewhat finer mesh:

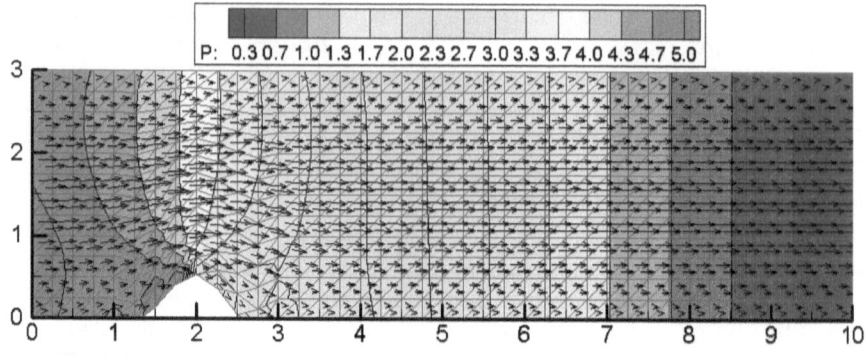

The last example we will consider (problem796) is yet another bump:

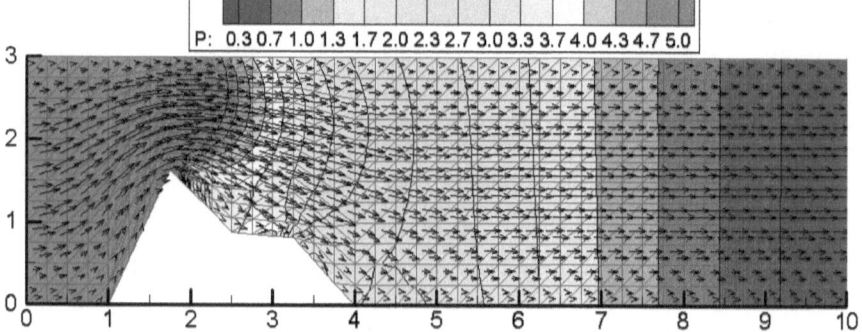

This program would be ever so much more useful had it been written in two parts: one to create the grid and a second to solve the equations.

3-Way Comparison

Of the three methods (FDM, FVM, and FEM), FDM is most intuitive, while FEM is least. The level of mathematics involved ranges from lowest (FDM) to highest (FEM). Difficulty of implementation at first glance follows the level of mathematics, but that's not the whole picture. If you ask researchers in the field of CFD which method is best or which is hardest to implement, you will get as many different answers as people you ask. I believe this lack of consensus arises less from mathematics and more from personal preferences regarding hard-to-quantify objectives like *tidiness* and *completeness*.

Preparing the input files for an FEM flow model can be quite tedious, which is why there are tools to do so associated with all of the popular codes, both commercial and open source. It is most convenient to specify the domain with a polygon, including the boundary conditions along the polygon, and using these to create the element/side specific boundary conditions for the FEM model to

70

read, after breaking the domain up into triangles. Several illustrations can be found in the online archive, including the following rather coarse model of a hypothetical lake:

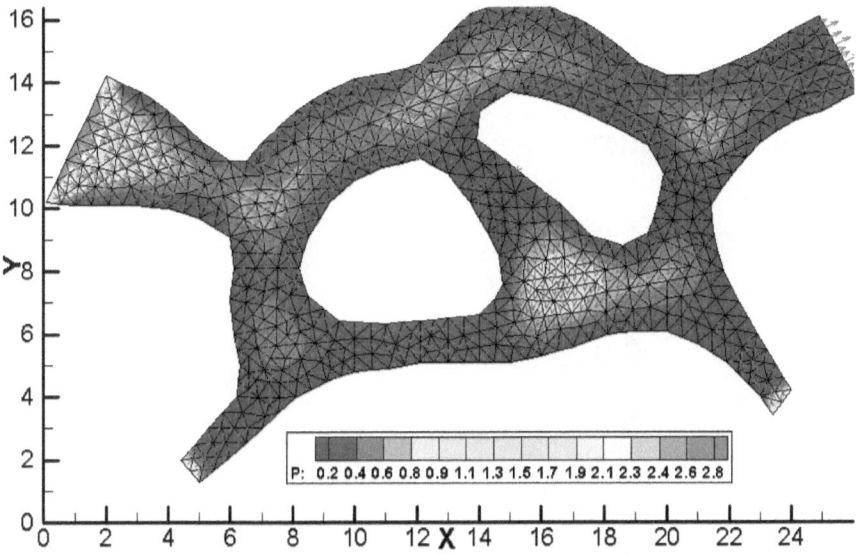

Pathlines represented by temperature-colored ribbons help visualization:

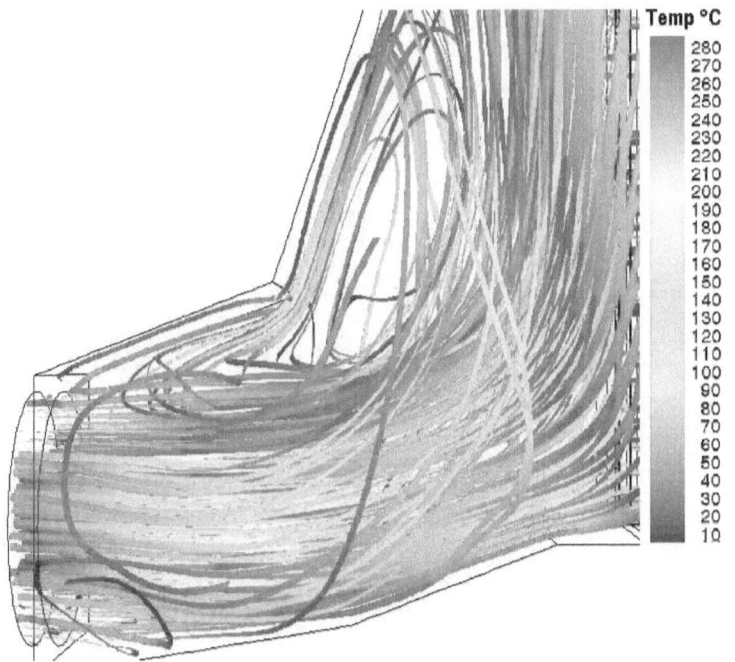

Flow and temperature within a reacting flow are illustrated in this figure:

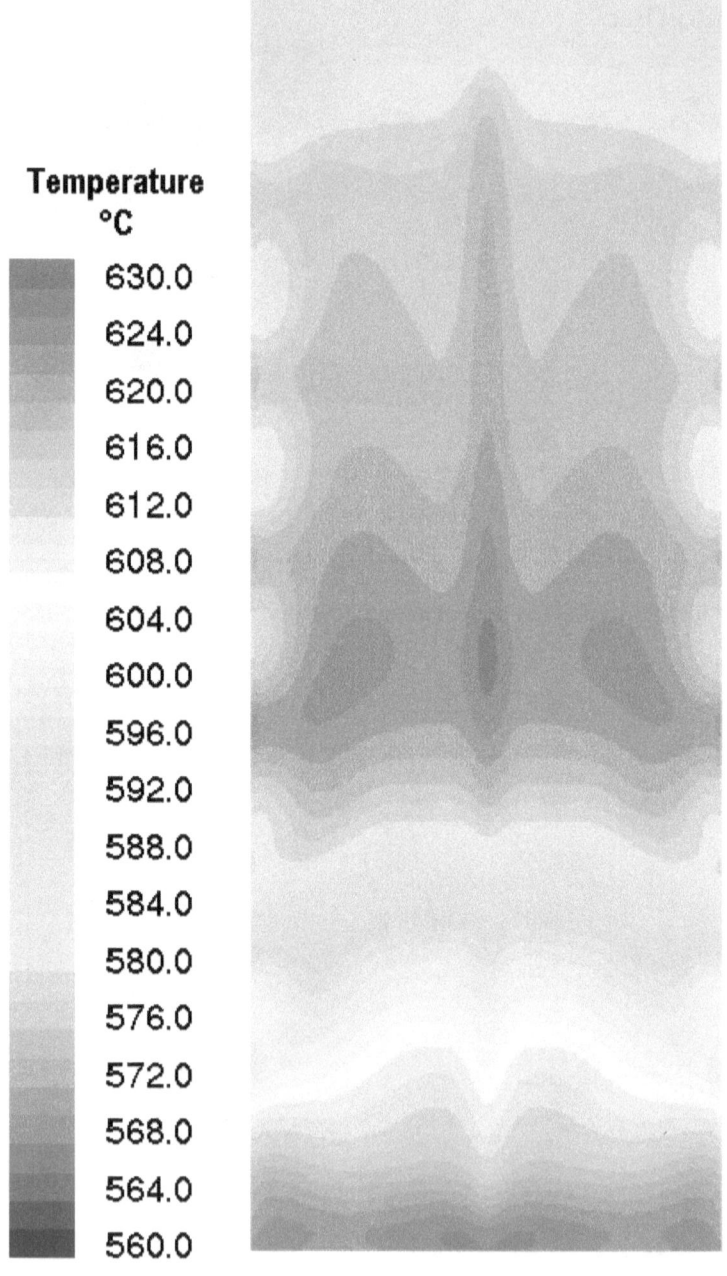

Chapter 6. Turbulence

Everyone who has experienced a blustery day knows something of turbulence. Air stumbling as it flows around a corner is commonplace, as is the seemingly chaotic rustling of autumn leaves. The observant notice a difference between slow and fast flows and that air and water are more unlike than merely density. While the difference between pouring honey and water is clear, a keen eye can distinguish between flowing water and alcohol. We intuitively understand that viscosity plays an important part.

Upon careful inspection, some flows or locations within a flow are turbulent, while others are not. Much effort has been applied to the study of this difference and how to predict the transition from laminar to turbulent. You should already be familiar with this concept, at least in pipe flow, where transition from laminar to turbulent starts at a Reynolds number of about 2000 and is complete at about 4000. Typical velocity profiles in a pipe are shown in the following figure:

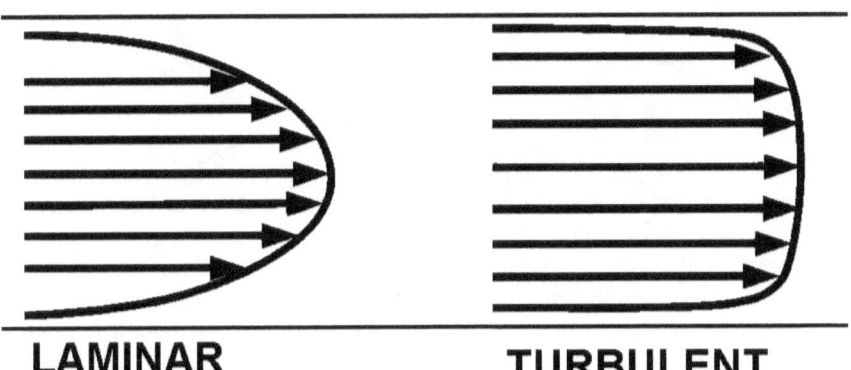

LAMINAR   TURBULENT

The laminar profile is parabolic (i.e., varies with r²) and the turbulent profile varies roughly with the one-seventh power of distance from the wall. The governing equations are the same (i.e., Navier-Stokes). The viscosity can't be constant across the radius and produce the profile on the right. Through empirical correlations (i.e., many measurements), we can predict what the effective (i.e., turbulent) viscosity must be to yield this profile. We can predict this fairly accurately for a few simple geometries (e.g., pipe, flat plate, tube bundle, etc.); however, there is no simple or foolproof way of predicting complex flows or arbitrary geometries. In spite of decades of research, there is still no analytical theory to predict the evolution of turbulent flows.

While there may be no foolproof way, there are some fairly reliable approximations. As we consider two of these methods, we must also note that the velocity profiles depicted above are time-averages and not instantaneous. While the laminar profile might not change perceptibly with time, the local turbulent velocity most definitely will. This next figure illustrates that to a somewhat exaggerated degree.

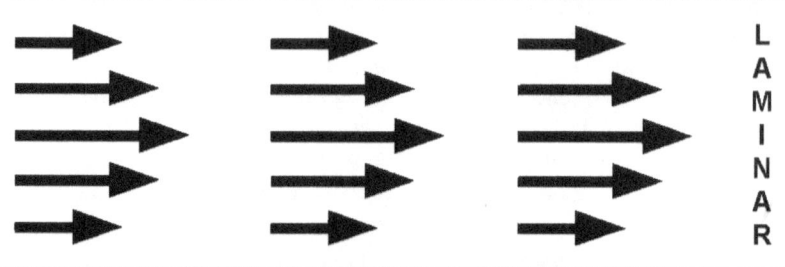

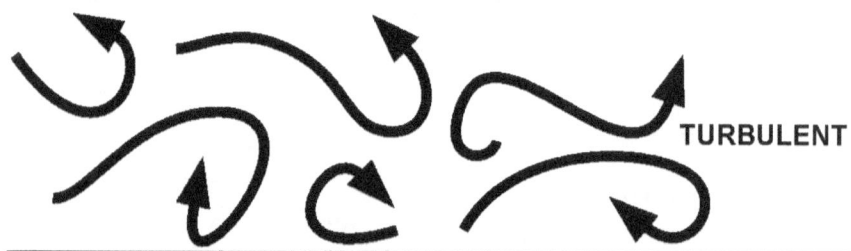

We must also realize that there is more than one time scale and more than one length scale operating within turbulent flows. Here we are interested with the small scale in both time and spatial fluctuations, not large scale or gross changes. I have found it most helpful to consider the ***mixing length hypothesis*** first introduced by Prandtl.[19] He envisioned little *blobs* of the fluid *lurching* about in the flow, participating in and contributing to the chaotic behavior we call turbulence. Two aspects of this concept fit with observations: 1) faster velocities are expected to have more blobs lurching more often; 2) blobs have more room to lurch in the open and less room to lurch when close to a boundary. The lurching blobs influence the flow as if changing its apparent properties or what we call *eddy viscosity*, as this behavior is easily seen in eddies.

All turbulent adjustment are attempts to capture and predict this behavior. These are necessarily based on empirical, that is, experimental measurements. They are also all approximations, as this is a transient phenomenon and has no fixed measurement, such as timing how long it takes for a liquid to drain out of a cup with a given diameter hole. We will consider only two of the many models.

---

[19] Ludwig Prandtl (1875–1953) German engineer and pioneer in the study of fluid flow.

k-ε (k-epsilon)

The predominant k-ε model was introduced by Chien and is the most common turbulence model used in CFD.[20] It is a two-equation model, which gives a general description of turbulence and is a marked improvement over the original mixing length concept. This model provides an algebraic expression that can be used in the computation of complex flows. The k-ε generally works better away from boundaries.

Turbulent dissipation is the rate at which velocity fluctuations dissipate and forms the basis of the k–ε model. The *eddy viscosity* is calculated from the average turbulence length scale, based on empirical coefficients. The *gradient diffusion hypothesis* relates the Reynolds stresses to the mean velocity gradients, yielding the effective turbulent viscosity. Not surprisingly, this calculation performs poorly near separation zones and strongly rotating flows.[21] It also does not account for strong adverse pressure gradients, which can occur for flows even without separation.[22]

All variants of k–ε are linear eddy viscosity models, based on the Boussinesq assumption:[23]

$$\tau_{ij} = 2\mu_t\left(S_{ij} - \frac{1}{3}\frac{\partial u_k}{\partial x_k}\delta_{ij}\right) - \frac{2}{3}\rho k \delta_{ij} \tag{6.1}$$

where τ_{ij} is the turbulent shear stress μ_t is the turbulent viscosity, x_k is any one of the principal dimensions, u_k is the corresponding velocity component, δ_{ij} is the Kronecker delta, k is a constant, and ρ is the density. S_{ij} is reminiscent of the vorticity and given by the following expression:

$$S_{ij} = \frac{1}{2}\left(\frac{\partial u_i}{\partial x_j} + \frac{\partial u_j}{\partial x_i}\right) \tag{6.2}$$

The turbulent eddy viscosity is computed from:

$$\mu_t = C_\mu f_\mu \frac{\rho k^2}{\varepsilon} \tag{6.3}$$

---

[20] Chien, K.-Y., "Predictions of Channel and Boundary-Layer Flows with a Low-Reynolds-Number Turbulence Model," AIAA Journal, Vol. 20, No. 1, pp. 33-38, 1982.

[21] Menter, F. R., "Zonal Two Equation k-w Turbulence Models for Aerodynamic Flows," AIAA Paper #93-2906, 24th Fluid Dynamics Conference, July 1993.

[22] Menter, F. R., "Two-Equation Eddy-Viscosity Turbulence Models for Engineering Applications," AIAA Journal, Vol. 32, No. 8, pp. 1598-1605, 1994.

[23] Joseph Valentin Boussinesq (1842–1929) French mathematician and physicist who made significant contributions to the theory of hydrodynamics, vibration, light, and heat.

At a boundary or wall, both **k** and ε are zero. The remaining equations are:

$$\sigma_k = 1.0$$
$$\sigma_\varepsilon = 1.3 \tag{6.4}$$
$$C_\mu = 0.09$$
$$C_{\varepsilon 1} = 1.35 \tag{6.5}$$
$$C_{\varepsilon 2} = 1.80$$
$$f_1 = 1$$
$$f_2 = 1 - \frac{2}{9} e^{-\frac{\text{Re}_T^2}{36}} \tag{6.6}$$
$$f_\mu = 1 - e^{-0.115 d^+}$$

$$\text{Re}_T = \frac{\rho k^2}{\mu \varepsilon} \tag{6.7}$$

$$d^+ = \frac{d \rho u_\tau}{\mu} \tag{6.8}$$

$$u_\tau = \sqrt{\frac{\tau_w}{\rho_w}} \tag{6.9}$$

$$\tau_w = \mu_w \left(\frac{\partial U}{\partial n} \right) \tag{6.10}$$

where the subscript, w, indicates *wall*. The parameter, d^+, is the dimensionless distance to the wall. The partial of U with respect to n is the rate of change of the velocity parallel to the wall with respect to the direction normal to the wall.

This nest of equations is daunting and many readers will wonder how they could possibly implement such a thing on top of all the already overwhelming Navier-Stokes equations. That's why I have included Mohammadi's NSC2KE program among the examples. You can browse through the source code (nsc2ke.for) and see how the turbulence model is implemented. In particular, study subroutines ke_law, ke_two, and log_law.

k–ω (k–omega)

The k–ω turbulence model introduced by Wilcox is also a two-equation turbulence model that is often used.[24] The model attempts to predict turbulence

---

[24] Wilcox, D. C., "Formulation of the k-omega Turbulence Model Revisited," AIAA Journal, 2008.

by two partial differential equations for two variables, k and ω, with the first variable being the turbulence kinetic energy (k) while the second (ω) is the specific rate of dissipation (of the turbulence kinetic energy k into internal thermal energy). The k–ω generally works better near boundaries, such as walls.

This model allows for a more accurate near wall treatment and switches from a wall function to a low-Reynolds number formulation based on grid spacing. The k–ω model performs better for wall-bounded and low Reynolds number flows and also better predicts transition. The k–ω model also performs much better with adverse pressure gradients. The model does, however, under-predict separation in severe adverse pressure gradient flows.

ISAAC

Integrated Solution Algorithm for Arbitrary Configurations (ISAAC) is a compressible Euler/Navier-Stokes computational fluid dynamics code. ISAAC includes the capability of calculating the Euler equations for inviscid flow or the Navier-Stokes equations for viscous flows. ISAAC uses a domain decomposition structure to accommodate complex physical configurations. ISAAC can calculate either steady state or time dependent flow.

ISAAC was specifically designed
to test turbulence models.

Various two equation turbulence models, explicit algebraic Reynolds stress models, and full differential Reynolds stress models are implemented in ISAAC. Several test cases are documented in the ISAAC was developed under contract to the NASA Langley Research Center while the author (Joseph H. Morrison) was employed by Analytical Services and Materials, Inc. ISAAC is public domain and so we utilize it here explore turbulence models.

You can find the manual plus all of the associated files in the online archive in folder examples\isaac. I have combined all of the many modules into a single source file and modified this to compile and run on Windows®. The output is written to files that can be displayed by Tecplot™ or TP2. I hope that in making available this version of ISAAC, which will run on 97% of the worlds' computers, this excellent program will be more widely used, especially by graduate students. You can experiment with the examples provided, find others on the web, and build your own.

There are five examples, each with a batch file and Tecplot™ layout. The first of these is ke, which is flow over a flat plate. The entire domain is:

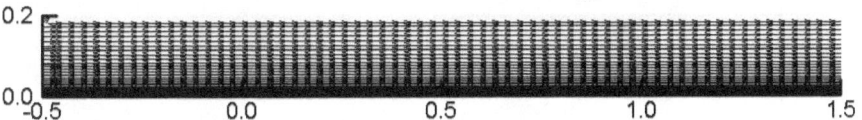

An expanded view near *x=0* is shown in this next figure:

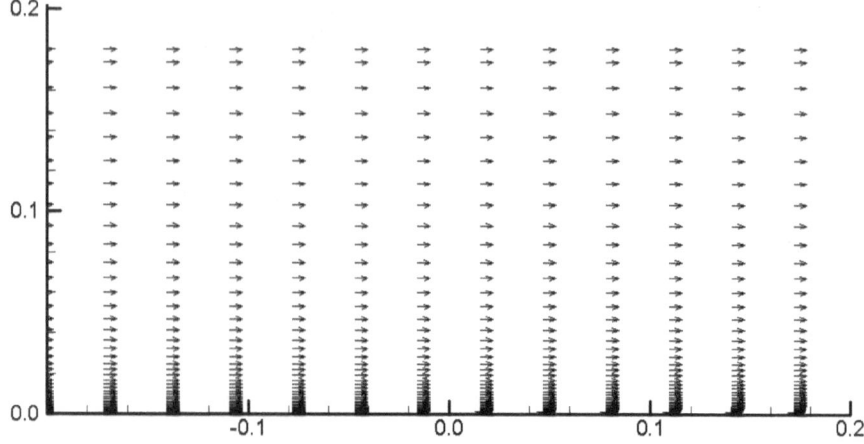

The NACA 0012 airfoil multi-grid example (naca0012_mg) domain is:

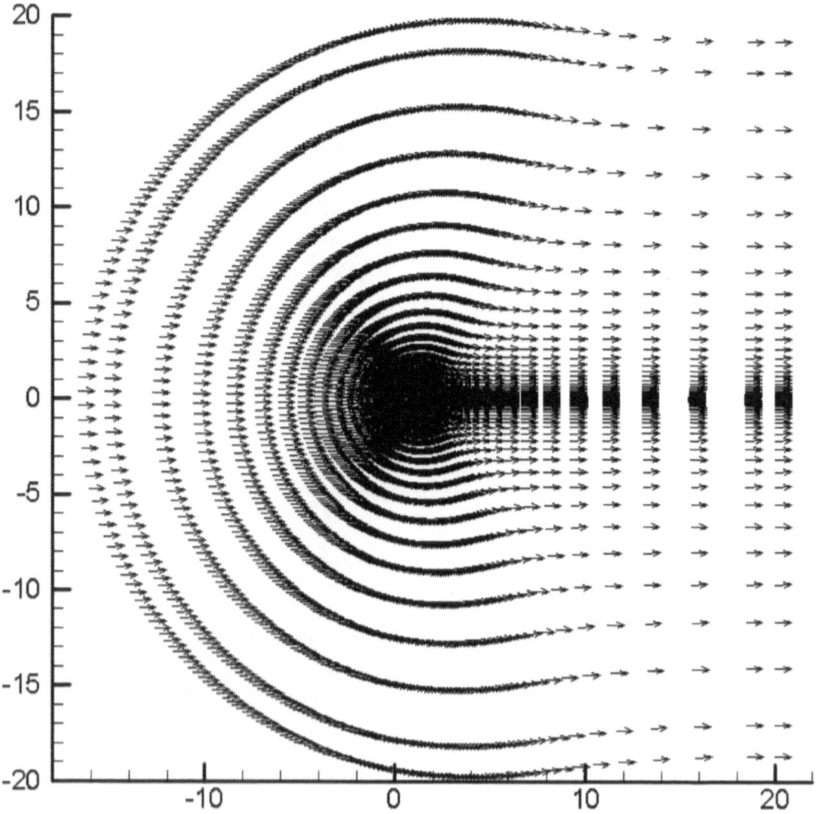

A closer look at the center:

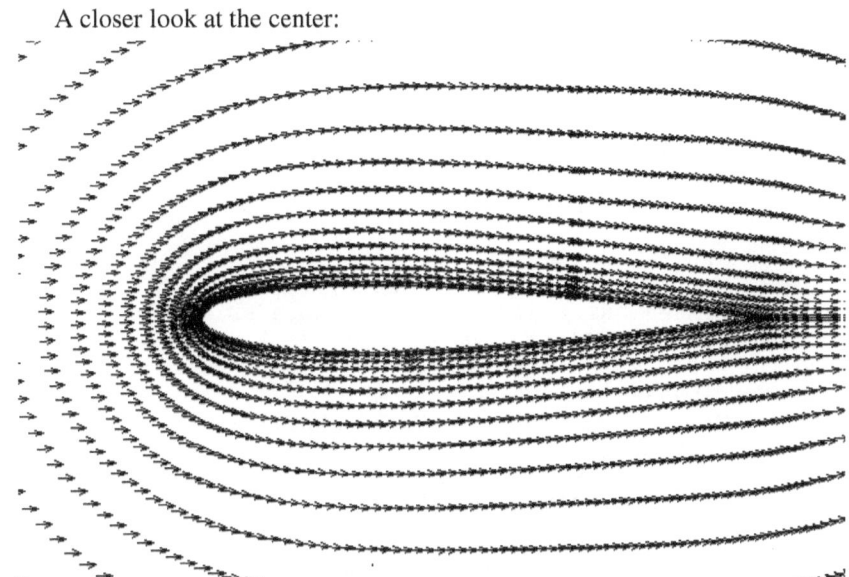

ISAAC.FOR is well documented and organized. Morrison uses many of the same symbols in the code, which makes it easy to see where the various turbulence models are implemented. For instance, the following section of code roughly corresponds to Equations 6.1 through 6.10:

```
C Set up normals for local coordinate system (s=SP,
C n=RN, t) where n is normal to the surface, s is
C parallel to the surface aligned with the flow, and
C t is perpendicular to s, n (and is not needed)
      RNX=S(JM,KM,IM,1,IDIR)*SGNM
      RNY=S(JM,KM,IM,2,IDIR)*SGNM
      RNZ=S(JM,KM,IM,3,IDIR)*SGNM
      UDOTN=Q(JN,KN,IN,2)*RNX+Q(JN,KN,IN,3)*RNY
     &+Q(JN,KN,IN,4)*RNZ
      UPN=Q(JN,KN,IN,2)-UDOTN*RNX
      VPN=Q(JN,KN,IN,3)-UDOTN*RNY
      WPN=Q(JN,KN,IN,4)-UDOTN*RNZ
      UPARN=SQRT(UPN*UPN+VPN*VPN+WPN*WPN)
      IF(UPARN.EQ.0.)THEN
        SPX=SHATX
        SPY=SHATY
        SPZ=SHATZ
      ELSE
        SPX=UPN/UPARN
        SPY=VPN/UPARN
        SPZ=WPN/UPARN
      ENDIF
      RHOW=Q(JBC,KBC,IBC,1)
```

```
        TAUW=PROPS(JBC2,KBC2,IBC2,ITQTAU)
        YNORML=ABS(PROPS(JN,KN,IN,4))
        YNORML=MAX(YNORML,RSMALL)
        RHO=Q(JN,KN,IN,1)
        TAUXX=Q(JN,KN,IN,6)
        TAUYY=Q(JN,KN,IN,7)
        TAUZZ=Q(JN,KN,IN,8)
        TAUXY=Q(JN,KN,IN,9)
        TAUXZ=Q(JN,KN,IN,10)
        TAUYZ=Q(JN,KN,IN,11)
        TKEN=0.5*(TAUXX+TAUYY+TAUZZ)
        UTAUW=SQRT(ABS(TAUW)/RHOW*FSMACH/RE)
        DUPDN=UTAUW/(RKAPVK*YNORML)
C Transform stresses to local streamwise, normal
C coordinates
        TAUNN=RNX*RNX*TAUXX+RNY*RNY*TAUYY+RNZ*RNZ*TAUZZ
       1+2.*(RNX*RNY*TAUXY+RNX*RNZ*TAUXZ+RNY*RNZ*TAUYZ)
        TAUSN=SPX*RNX*TAUXX+SPY*RNY*TAUYY+SPZ*RNZ*TAUZZ
       1+(SPX*RNY+SPY*RNX)*TAUXY+(SPX*RNZ+SPZ*RNX)*TAUXZ
       2+(SPY*RNZ+SPZ*RNY)*TAUYZ
```

This next section of code is where the constants are set depending on the turbulence model:

```
C Read in values of constants
C
        DO ICON=1,NCON
          READ(INPUT,*,END=990,ERR=995)VAR,VALUE
          IF(VAR.EQ.'SAA')THEN
            IEPSEQ=IEESAA
            IDAMP=IFMSAA
            IFDL2K=0
            SIGK=1.0/1.00
            SIGT2=1.0/1.36
            SIGRHO=1.0/0.95
            CMU=0.09
            CEPS1=1.44
            CEPS2=1.83
            A2KE=4.9
            ALF1=0.6
            ALF2=0.4
            ALF3=0.2
            CPDFRO=0.
            CPDFT=0.
            CPVELK=0.
          ELSEIF(VAR.EQ.'ZSGS')THEN
            IEPSEQ=IEEZSG
            IDAMP=IFMZSG
            IFDL2K=1
            SIGK=1.0/1.00
```

```
      SIGT2=1.0/1.45
      SIGRHO=1.0/0.95
      CMU=0.096
      CEPS1=1.50
      CEPS2=1.83
      A2KE=4.9
      ALF1=0.0
      ALF2=0.0
      ALF3=0.0
      CPDFRO=0.
      CPDFT=0.
      CPVELK=0.
    ELSEIF(VAR.EQ.'ZSSL')THEN
      IEPSEQ=IEEZSL
      IDAMP=IFMZSG
      IFDL2K=1
      SIGK=1.0/0.75
      SIGT2=1.0/1.45
      SIGRHO=1.0/0.50
      CMU=0.096
      CEPS1=1.50
      CEPS2=1.83
      A2KE=4.9
      ALF1=0.0
      ALF2=0.0
      ALF3=0.0
      CPDFRO=0.
      CPDFT=0.
      CPVELK=0.
    ELSEIF(VAR.EQ.'HIGHRE')THEN
      IEPSEQ=IEEHR
      IDAMP=IFMHR
      IFDL2K=0
      SIGK=1.0/1.00
      SIGT2=1.0/1.30
      SIGRHO=1.0/0.95
      CMU=0.09
      CEPS1=1.44
      CEPS2=1.92
      A2KE=4.9
      ALF1=0.0
      ALF2=0.0
      ALF3=0.0
      CPDFRO=0.
      CPDFT=0.
      CPVELK=0.
    ELSEIF(VAR.EQ.'RNG')THEN
      IEPSEQ=IEERNG
      IDAMP=IFMHR
```

```
        IFDL2K=0
        SIGK=1.39
        SIGT2=1.39
        SIGRHO=1.0/0.95
        CMU=0.085
        CEPS1=1.42
        CEPS2=1.68
        A2KE=4.9
        ALF1=0.0
        ALF2=0.0
        ALF3=0.0
        CPDFRO=0.
        CPDFT=0.
        CPVELK=0.
      ELSEIF(VAR.EQ.'ASM')THEN
        IASM=IASMGS
        ICMUST=5
        IEPSEQ=IEEABD
        IDAMP=IFMHR
        IFDL2K=0
        SIGK=1.0/1.00
        SIGRHO=0.0
        CMU=0.081
        CEPS1=1.44
        CEPS2=1.83
        A2KE=5.5
        ALF1=0.0
        ALF2=0.0
        ALF3=0.0
        CPDFRO=0.
        CPDFT=0.
        CPVELK=0.
        SIGT2=1.0/(0.41**2/((CEPS2-CEPS1)*SQRT(CMU)))
      ELSEIF(VAR.EQ.'ADRM')THEN
        IADRM=IADRGS
        IEPSEQ=IEEHR
        IDAMP=IFMHR
        IFDL2K=0
        SIGK=1.0/1.00
        SIGT2=1.0/1.87
        SIGRHO=0.0
        CMU=0.094
        CEPS1=1.20
        CEPS2=1.83
        A2KE=0.0
        ALF1=0.0
        ALF2=0.0
        ALF3=0.0
        CPDFRO=0.
```

```
        CPDFT=0.
        CPVELK=0.
      ENDIF
    ENDDO
```

This next section shows that ISAAC handles the k-ω model:

```
C   k-omega constants from Wilcox, AIAA Journal, Vol.
    26, No. 11
      ELSEIF(VAR.EQ.'KOMEGA')THEN
        ITURB=ITKW
        IEPSEQ=0
        IDAMP=IFMHR
        IADRM=IADRNO
        IASM=IASMBU
        IEPSC=IECNO
        IPDIL=IPDNO
        ICMUST=0
        IEPSLN=7
        IFDL2K=0
        NQ=7
        NP=5
        NF=7
        NRANK=7
        SIGK=0.5
        SIGT2=0.5
        SIGRHO=0.0
        BSTRKW=0.09
        BKW=3.0/40.0
        CMU=1.0
        GKW=5.0/9.0
        ROUGHK=0.0
        CPDFRO=0.0
        CPDFT=0.0
        CPVELK=0.0
        PRDLIM=100.
```

Chapter 7. Three Dimensional Flow

Effectively managing 3D elements and solving models is a daunting task, requiring years of experience and very complex code—far more than we can cover here. I direct you to the many online resources. Commercial codes, such as Fluent™, PHOENICS™, and AutoDesk™ CFD, are out of reach for many; however, there are many free and open code solvers and grid creation tools, such as Open FOAM. One or more of these may be quite adequate for your particular needs. There are several comparisons on the Web, for instance, the following, which is free of commercialism:

https://www.cfd-online.com/Wiki/Codes

Below is another list you may find helpful:

http://www.cfdyna.com/Home/CFD_Softwares.html

Do not expect to find some free and easy-to-use suite that will enable you to rapidly create and solve meaningful models. This is a long and tedious task even with the best of tools. Expect to devote considerable time to mastering whatever tools you select. Unless you plan to devote your entire career to this task, you will need to rely heavily on the work of others.

Appendix A. Tecplot™ vs. TP2

Visualization is a key part of CFD. We are often more interested in flow patterns than specific values of pressure and velocity. Tecplot™ is a powerful and versatile tool for visualization of many different types of data, but particularly fluid flow, as this software arose from and was motivated by early CFD research. The developers of this software had connections with NASA and were deeply involved with aerodynamics. This excellent product can be found at their web site: https://www.tecplot.com/

I developed TPLOT in 1980 while working on my doctorate to graphically display the data I was collecting in the laboratory. It originally only worked on one device: Tektronix 4010. That's where the "T" in TPLOT came from.

Over the years I added many devices and continued to use TPLOT as I worked in industry. TPLOT was written in FORTRAN, which became increasingly problematic, as operating systems evolved. In the summer of 1993, I began work on the second generation of TPLOT, which I named TP2. This new code was written in C, which opened up many more devices, but it was still not technically a Windows® application. That change didn't come until the spring of 1998. There have been dozens of revisions and additions since then. TP2 is available free online:

https://dudleybenton.altervista.org/software/index.html

Perhaps the biggest difference between Tecplot™ and TP2 is file types. The file extension doesn't mean anything to Tecplot™. That is, the same data could be stored in a file with any extension. For Tecplot™ the data structure is defined by various headers and there can be multiple types of data in the same file. For TP2 the data structure is identified by the file extension, as indicated below, and there can be only one data structure in a file.

filename.extention

In order to control how the data is presented in Tecplot™ you also need a layout file (usually, but not necessarily, filename.lay). TP2 displays data based on the type, which is indicated by the file extension. Tecplot™ will display 3D data as 3D or 2D with or without contours, shading, slicing, etc. TP2 will always display 3D data in 3D, unless you are specifically slicing at a plane. TP2 recognizes many more data structures than Tecplot™ (27 in all), each one having a different file extension. TP2 also has a layout (file extension TP2) and can display data from multiple files and of multiple structures, but this isn't mandatory, as with Tecplot™.

Surfaces and Volumes

Both Tecplot™ and TP2 handle 2D surfaces and 3D volumes. The display is similar, but the data structures are different. Tecplot™ requires the file to contain every x,y,z for 2D or x,y,z,w for 3D. One of the main motivations for developing TP2 was efficiency and compactness, including the smallest possible file sizes. If the surface or volume is complete (i.e., rectangular), whether or not the spacing is uniform, there is a definite pattern so that it is superfluous to enter all of the points except for z in 2D and w in 3D. That's how TP2 works. Surfaces are defined by a 2D table (file extension TB2) and volumes are defined by a 3D table (file extension TB3). The data are entered as a list of x's, then y's, then z's, then w's—not each and every x,y,z,w.

Finite Elements

Finite elements (triangles, quadrangles, tetrahedra, bricks, etc.) are very similar in Tecplot™ and TP2. The file structure consists of a list of nodes followed by a list of element indices. For TP2 2D finite elements have a file extension of 2DV and 3D finite elements have the extension 3DV.

Velocity Vectors

Velocity vectors are handled differently. In Tecplot™ the velocity components (typically u, v, and w) can be in any column, but are defined along with spatial coordinates (x, y, and z). With TP2 2D velocity vectors consist of x, y, u, and v—in that order and in a file having the extension V2D. Three-dimensional velocity vectors consist of x, y, z, u, v, and w—in that order and in a file having the extension V3D.

Layout

As mentioned before, Tecplot™ *requires* a layout file, which usually has the extension LAY. TP2 accepts an *optional* layout file, which has the extension TP2. With TP2, you can override any file extension (for example reading 2D velocity vectors from a file with extension VEC) by appending a minus followed by the intent, as in:

TP2 velocities.vec-v2d

Of course, this means that with TP2, you can't plot data from files that have a minus contained in the name, as this will be interpreted, truncating the file name.

Multiple Document Interface

Windows® recognizes what is called a *multiple document interface*, or MDI. TP2 is based around this concept, while Tecplot™ is not. Tecplot™ will only display a single context. TP2 will display up to 25 completely unrelated contexts, each in its own window. TP2 can be launched with wild cards:

TP2 *.v2d *.v3d *.2dv *.3dv

Animations

Tecplot™ will create animations—raster meta files (extension RM) for early versions and also audio visual interleave files (extension AVI) for recent versions. A utility, Framer, is provided with Tecplot™ to display the animations so created. AVI files can be displayed by various utilities. TP2 creates and also displays various animations in several formats, including GIF.

Examples

Tecplot™ comes with several excellent examples. These are separate files in a subfolder created during installation. TP2 comes with a variety of 2D and 3D examples, all of which are embedded inside the executable, so that you only need the EXE file with TP2. When you select a demo, TP2 creates the files and then displays the results. Both programs come with help files.

Data Processing

Both Tecplot™ and TP2 process data in a variety of ways, including interpolation, cutting, slicing, and translation from one form to another. TP2 has far more options for this because I added a feature every time I needed one for the work I was doing.

Drawings and Objects

TP2 will also read some AutoCAD™ files, including DXF and 3DS (3D Studio) as well as virtual reality markup language (VRML) files, which Tecplot™ will not import.

Appendix B. Compilers

If you don't have a C compiler, I suggest either Digital Mars® or Microsoft®. The former can be downloaded free from the following link:

http://www.digitalmars.com/

The Microsoft® C compiler is also available free of charge. The thing that costs so much is the Visual Studio® Interactive Development Environment (IDE), which is completely unnecessary and extremely annoying. Simply download and install the W7.1 SDK and DDK. While these developer kits are no longer available at the Microsoft@ Download Center, they can be found elsewhere on the web.

After you install the two kits, combine the bin, include, and lib folders and put them in a folder called something like C:\VC32 or C:\VC64. There will be several folders with similar names. The folders you need will have either x86 or x64 in them. There are four combinations of the two architectures. These arise from the O/S you are running them on and the O/S you are targeting. For instance, you can create a 64-bit executable on a 32-bit machine and vise versa. Unless you need more than 2GB of memory in a single program, 64-bit is not necessary, as 32-bit executables will run on either O/S. You will, however, need to create the specific target when creating Add-Ins for Excel, as these are not interchangeable.

If you can tolerate the nagging social media stuff, dreadful user interface, and every possible wrong default setting (like unicode and .Net targeting), you can get the Visual Studio® Express Community version for free at this link:

https://visualstudio.microsoft.com/vs/express/

I do not recommend the Gnu® compiler, gcc, as it was developed specifically for Linux®. While it does run on Windows® and may produce viable executables, it is very quirky, assumes all sorts of things that aren't ANSI (as far as I can tell), and is nothing but trouble. Considering that both the Digital Mars® and Microsoft® compilers are available free, there is no reason to use gcc, unless you're stuck with Linux®.

The Intel™ C compiler touts extended features and convenient access to unlock the full power of their processors. I used it extensively at one time, but now see no advantage to it. The early Microsoft® C compilers were a dreadful mess of bugs and would croak if you turned on any of the optimization options. The Microsoft® FORTRAN compilers (including Power Station™) before the now defunct Digital Equipment Corp. (DEC) and later HP™/Compaq™ fixed it were also dreadful. Sometime around 2005 Microsoft® must have given the task of fixing their C compiler to an entirely new group. [I'd love to meet them, for they did a splendid job!] The 2005 (XP3) and 2010 (W7) Microsoft C compilers work quite well, precluding any need I've had for the Intel™ C compiler.

https://software.intel.com/en-us/c-compilers

FORTRAN

I once worked on a project for the U.S. Departments of Energy and Defense (DoE and DoD) to parallelize a FORTRAN flow code that was used extensively on various projects. We devoted much effort to employ Portland Group's compiler. This aspect of the project was very disappointing and withered for lack of continued funding. One of the team members also worked for the High Performance Computing (HPC) Group at Oak Ridge National Laboratory (ORNL), which has been very successful; so our lack of success was not due to a paucity of talent. In short, that compiler was not worth the investment. This likely had more to do with the difficulty of parallelizing the code than any deficiencies of the compiler. It seems that many efforts of the HPC are focused on finding applications that can be effectively parallelized rather than figuring out how to parallelize applications that are needed.

During that time we also tested the now defunct Watcom C and FORTRAN compilers. These are still available on the web, though neither is necessary, considering the other options available. When needed, I still use V6.1 of DEC's Digital FORTRAN and also keep a copy of Microsoft's last 16-bit C and FORTRAN compilers for legacy applications. There is also a Gnu FORTRAN compiler available for Linux. I have found that much of the *free* FORTRAN code available on the web will *only* work with the Gnu compiler, as these codes contain statements like: *use* this, that, or the other, which refer to namespaces that never existed on a DEC™ mainframe, so they're not part of DEC's FORTRAN, which is the reigning standard.

I began writing code in 1972 using FORTRAN, and soon after that assembler, because there were so many things that couldn't be done with FORTRAN in the early days. Over the years, I've written a million lines of FORTRAN and assembler plus another two million in C, so I speak from experience. FORTRAN, like BASIC, is a convenient way of throwing something together quickly. In November of 1988, I was invited to give a lecture on the *Future of Computers in Engineering*. It is remarkable how many of those predictions have come true in the past 31 years. FORTRAN should have faded away long ago like Pascal and rotary phones. BASIC would have, were it not for the ease of creating interactive Windows® applications with Visual BASIC, compared to the painful path of C to accomplish similar results. Even now, Visual BASIC is morphing into C# so that soon you won't be able to tell the difference.

Python

I often correspond with scientists and engineers who develop python code. If you can write python, then you can write C. Every library function you could ever want in python or MATLAB® can be found in C on the Web. Python compilers and MATLAB® are superfluous when you consider how many C compilers are freely available. Using a system that automatically allocates and initializes variables for you and conveniently handles array overruns and divides by zero facilitates the entrenchment of bad habits. The Domain of Excellence lies along the Path of Discipline. Ditch python and learn C!

Code Profiling

The Intel® compilers provide runtime profiling: function calls and timing, though somewhat cumbersome and expensive. In over 40 years I've never seen anything that comes close to Walter Bright's C compilers (originally Zortech, then Symantec, and now Digital Mars). Simply add –gt and recompile to get a list of who called what, when, how many times, and how long each took. The convenience and simplicity is in a league of it's own—plus it's free! Walter Bright is a genius!

Appendix C: Finite Differences

Perhaps the most useful and complete reference on mathematical functions ever written is the *Handbook of Mathematical Functions* by Abramowitz and Stegun. It was first published by the National Bureau of Standards as Technical Monograph No. 55. The entire text is available free on-line at several locations, including:

http://people.math.sfu.ca/~cbm/aands/abramowitz_and_stegun.pdf

I can't say enough about this excellent text. Every serious student of mathematics must have a copy. You can also get it in paperback from several places, including E-Bay. This is what the cover looks like. If you happen to come across a copy in a used bookstore, buy it!

Chapter 25, Numerical Interpolation, Differentiation, and Integration by Philip J. Davis and Ivan Polonsky, contains an excellent section on finite differences.

Appendix D. Arrays

If you search the web for CFD codes, you may be surprised to see how many of these are still written in FORTRAN—a language associated with mainframes. It's also surprising to see how many arrays are statically allocated, in spite of the fact that dynamic allocation was introduced with F90, almost 30 years ago. While it may be convenient to compose code in FORTRAN and let the compiler worry about how to implement it, this is inefficient for processing. Admittedly, FORTRAN compilers are good at translating formulas and even multi-dimensional arrays, but the I/O is still archaic.

The C language works very much like assembler, which is the machine code of processors. FORTRAN is quite different. One of the more important differences between these two languages is how they handle arrays. Two, three, or even four-dimensional arrays in FORTRAN are not implemented as such. These are one-dimensional, contiguous blocks of memory. The FORTRAN compiler converts multi-dimensional addresses into a single offset, that is, a one-dimensional array. There are many C codes on the Web in which multi-dimensional FORTRAN arrays have been converted into C. These translations allocate two-dimensional arrays as pointers to arrays and three-dimensional arrays as pointers to pointers to arrays. If the developers could see the snarl of machine language that such a maneuver generates, they wouldn't do this.

I have written hundreds of thousands of lines of assembler for three different processors, including Intel®, so I know how this works. The following lines of FORTRAN or C:

```
X(I)=0.5
x[i]=0.5;
```

produce the same machine instructions, which you can see if you list the assembler or disassemble the executable.

```
MOV EAX,DWORD PTR DS:[BP+2]
MOV ECX,DWORD PTR DS:[BP+6]
FLD QWORD PTR DS:[EBP+10]
FSTP QWORD PTR DS:[ECX+EAX*8]
```

The first line loads the integer i into register EAX. The second line loads the beginning of array x[0] into register ECX. The third line loads a 64-bit (qword) floating point constant (0.5) onto the FPU stack, which converts it to 80-bit (tbyte), whether you're going to use it that way or not, because Intel® FPUs work with ten-byte floating point numbers. The fourth line converts the 80-bit floating point number to a 64-bit one, stores it in the location DS:[ECX+EAX*8], and pops the number off the FPU stack, discarding it.

The address x[i] consists of the beginning of x plus 8 times the value of i. The pitch of x (size of each element) is 8, because it's an array of 64-bit floating-point numbers. The preprocessor takes EAX, shifts it to the left 3 bits (multiplies it by 8), adds it to ECX, changes the instruction to DS:[ECX], and passes it along to the main processor. If the processor were operating in 16-bit or 32-bit *real* mode (i.e., DOS® or DOSX), we would almost be done, but not if it's running in 32-bit or 64-bit *protected* mode. In *real* mode, the ES, DS, SS, and CS registers are just numbers that refer to a block of memory. This is first shifted left 4 bits (multiply by 16) and added to the offset (DI, SI, BP, or IP) to form a 20-bit address, which is why DOS® was limited to 1 MB.

In *protected* mode, which is how all versions of Windows® since 3 run, the segment part (ES, DS, CS, SS, FS, or GS) isn't a segment at all; rather it's an index into a table that contains *handles* of blocks of protected memory, assigned by the O/S to your program. The processor must now go into the table and get the handle, then convert this to a literal address, convert it from block and offset to a physical address, load that onto the memory buss, notify the O/S, and wait for the result. All of that is necessary to keep you from crashing some other program, including the O/S itself.

We're still not done... The O/S must verify your privilege level and ownership of the requested memory block, which means more lists to check. If you launch another program from your program, informing the O/S that this is a debug operation, you can get at the memory belonging to the child process (i.e., step over the protective fence because you own it). This is how debuggers work and how hackers get into your code and mess around. The O/S provides this wonderful functionality, but at a cost. The bottom line is: there's a lot of overhead associated with each address. You do not want to use pointers to pointers unless it's absolutely necessary, which it most assuredly is not for something as simple as a 2D or 3D array.

Appendix E. Green's Lemma

Green's Lemma[25] is usually covered in advanced calculus. This useful theorem transforms an area (2D) integral into a boundary integral (1D). It is often used in developing finite element solutions. The Lemma can be expressed by the following integral:

$$\iint \left(\frac{\partial f}{\partial x} - \frac{\partial g}{\partial y} \right) dxdy = \oint \left(fdx + gdy \right) \tag{E.1}$$

While it can quite complicated to integrate a function over an irregular area (e.g., finite elements), it may not be nearly as complicated to integrate around a simple boundary. The spanning or basis functions used in finite elements are often simple and can be analytically integrated over one or the other dimension. This step provides the transformation from $\partial f/\partial x \to f(x,y)$ and $\partial g/\partial y = g(x,y)$.

Integrating around a boundary (most often a polygon) is quite simple. Furthermore, this often doesn't require a high degree of quadrature. In the case of finite element basis functions, 4-point Gauss Quadrature is often adequate. For illustration, we will integrate the basis function $X^a Y^b$. Below is a familiar polygon that we will use as an example:

The code (lemma.c) can be found in the online archive in examples\lemma. We use Gauss Quadrature to integrate 1D and 2D to illustrate that Green's Lemma works as it's supposed to. The code loops through values of a and b to produce:

```
illustrating Green's Lemma
  a    b    1D    2D
0.0  0.0   3.9   3.9
0.0  0.5   2.6   2.6
0.0  1.0   2.0   2.0
0.0  1.5   1.6   1.6
0.0  2.0   1.3   1.3
etc.
3.0  0.0  96.4  96.4
3.0  0.5  71.8  71.8
3.0  3.0  31.6  31.6
```

---

[25] George Green (1729-1841): British mathematical physicist best known for work with electric fields and magnetism.

There is also a test in the fem2d folder that includes the traditional polygon area calculation (which is based on Green's Lemma), 2D Gauss Quadrature, and Green's Lemma integrating with respect to *x* and with respect to *y* to show that it doesn't matter. Triangles are created from three random nodes:

```
poly[0].x=randbetween(-5000,5000)/1000.;
poly[0].y=randbetween(-5000,5000)/1000.;
poly[1].x=randbetween(-5000,5000)/1000.;
poly[1].y=randbetween(-5000,5000)/1000.;
poly[2].x=randbetween(-5000,5000)/1000.;
poly[2].y=randbetween(-5000,5000)/1000.;
if(PolygonArea(poly,3)<0.)
  {
  POLY p;
  p=poly[1];
  poly[1]=poly[2];
  poly[2]=p;
  }
for(j=0;j<3;j++)
  {
  ylop[j].x=poly[j].y;
  ylop[j].y=poly[j].x;
  }
printf("%6.3lf %6.3lf %6.3lf %6.3lf\n",
  PolygonArea(poly,3),Q(),
  GreensLemma(poly,3,f1),
  GreensLemma(ylop,3,f2));
```

The points are stored in the structure *poly*. If the polygon isn't ordered clockwise, points 2 and 3 are swapped. The same points are stored in *ylop* with *x* and *y* swapped. Then the integrations are performed to demonstrate that the same result is obtained either way.

```
testing integrations
PolyA   GQ2D LemaX LemaY
 0.35   0.35  0.35  0.35
 7.02   7.74  7.02  7.02
17.22  17.22 17.22 17.22
 6.06   6.05  6.06  6.06
 3.07   4.43  3.07  3.07
 3.11   3.10  3.11  3.11
 7.16   8.30  7.16  7.16
15.17  15.14 15.17 15.17
 7.73   7.72  7.73  7.73
34.82  34.83 34.82 34.82
```

Appendix F. Elements

Several errors can creep into a finite element model, including: unused nodes, coincident boundaries, overlapping boundaries, coincident nodes, elements having duplicate nodes, and degenerate elements. The program (check2d.c) in folder examples\check2d checks for all of these errors. Typical inputs are:

```
51 nodes
421.333 733.333
460.667 854.667
500 976
etc...
70 elements
2 3 4
8 9 10
26 27 28
etc...
```

Typical outputs are:

```
CHECK2D: check 2D elements
input file=star.2dv
reading nodes
  expecting 51 nodes
  allocating memory for nodes
  51 nodes found
  0≤X≤1000
  24≤Y≤976
reading elements
  expecting 70 elements
  allocating memory for elements
  70 elements found
checking for errors
  unused nodes... none
  coincident boundaries... none
  overlapping boundaries... none
  coincident nodes... none
  elements having duplicate nodes... none
  degenerate elements... none
elements are clockwise
```

This code contains several useful functions, including: polygon area and inside polygon test.

Grid Generation

You will also find a simple triangular mesh generator (elem3.c) in folder examples\elem3. It takes a bounding polygon and breaks the region into triangles. Several examples (lid, ring, sector) show how to form inclusions (holes inside the domain). Grid refinement is controlled by an optional parameter, as in:

101

```
elem3 lid.p2d 0.1
```

The first parameter (lid.p2d) is the bounding polygon file. The second (optional) parameter is the characteristic length, which depends on the scale of the domain.

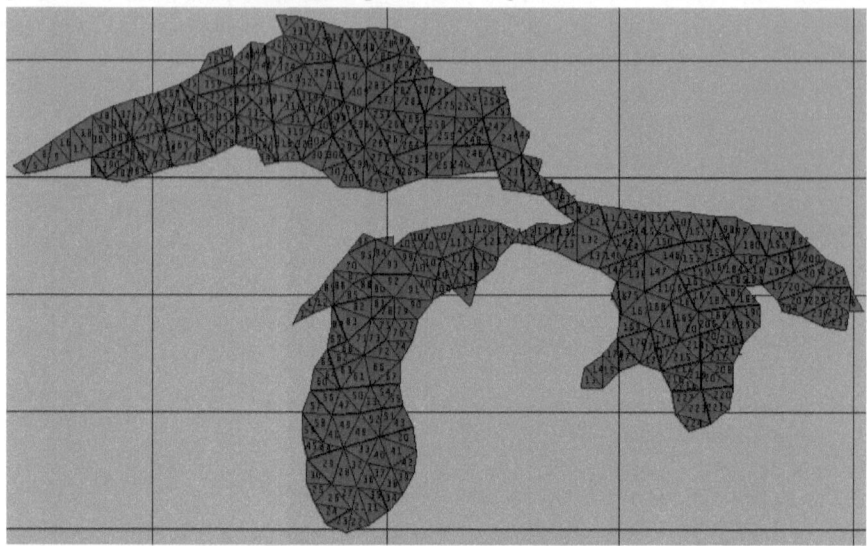

results of elem3 lakes.p2d

Appendix G: Exact Solutions

There are a few exact solutions to the Navier-Stokes equations. These have been programmed and can be found on Burkardt's web site. Search for "exact" on this webpage:

http://people.math.sc.edu/Burkardt/f_src/f_src.html

The 2D and 3D codes (exact2d.c and exact3d.c) have been translated and modified for the Microsoft C compiler and can be found in the online archive in folder examples\exact. One example is shown below:

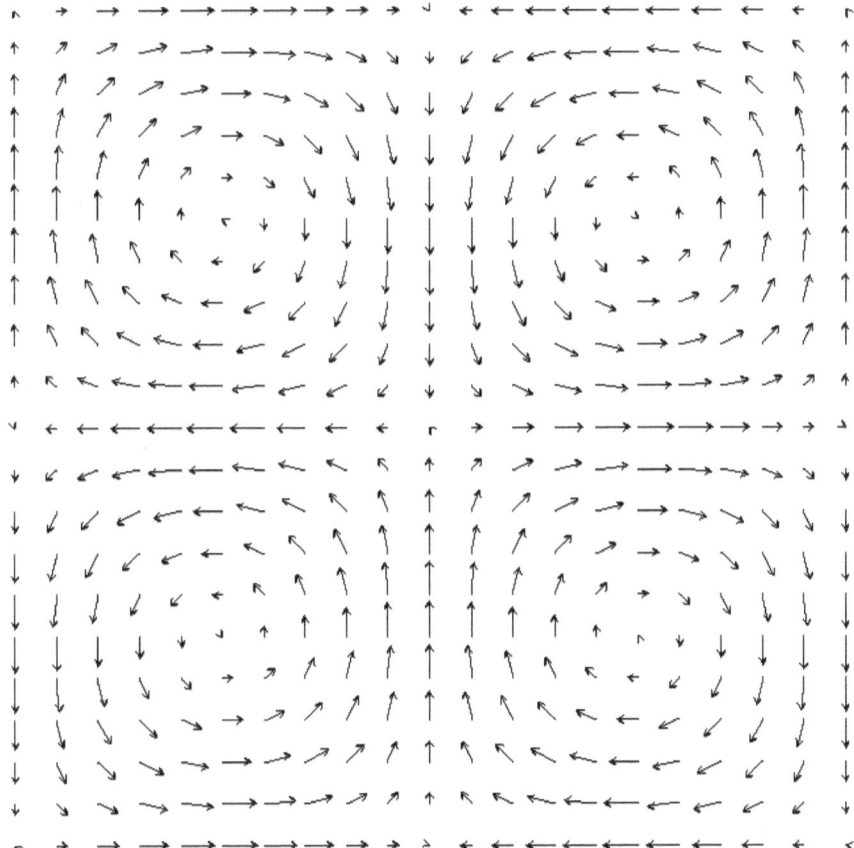

examples\exact\exact2.c with option TAYLOR

Appendix H. Transient Solutions

While we are not focusing on transients in this text, many problems have crucial transient aspects. While there are many FEM codes that calculate transients, this is not efficient. As illustrated in Chapter 3, it is far more efficient to employ finite differences and simply crunch along, stepping through time. There is a very interesting implementation of this process presented by Griebel, Dornseifer, and Neunhoeffer.[26] This approach was modified by D. Orchard and the code can be found on GetHub:

https://github.com/dorchard/navier/blob/master/c/nast2d-dom/init.c

The original code has been streamlined considerably and modified to create output files that are recognized by ubiquitous operating systems (i.e., Windows® and not LINUX). The code (nast2d.c) and associated files can be found in the online archive in folder examples\nast2d. The original license and readme file can also be found in this folder. Three snapshots are depicted on page iv. The code can be modified to change the obstructions, as illustrated below:

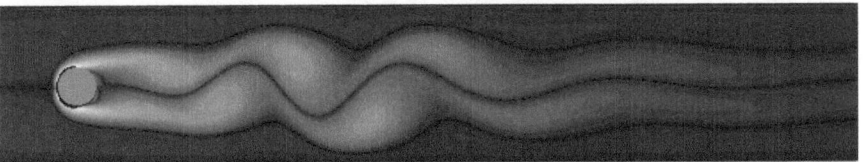

examples\nast2d\nast2d.c

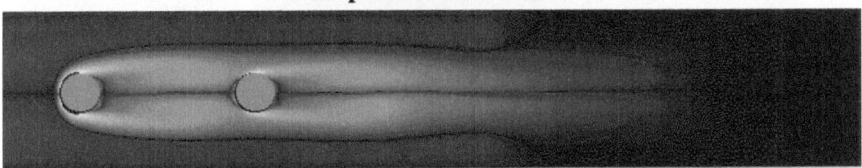

examples\nast2d\nast2d.c with option MAKE_TWO

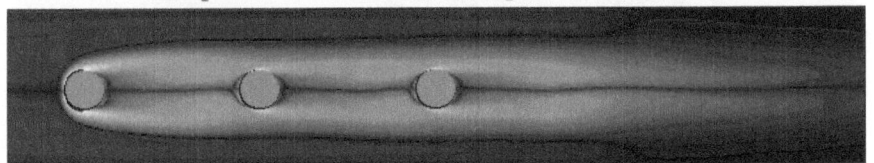

examples\nast2d\nast2d.c with option MAKE_THREE

---

[26] Griebel, M., Dornseifer, T., and Neunhoeffer, T., "Numerical Simulation in Fluid Dynamics," SIAM, 1998.

examples\nast2d\nast2d.c with option MAKE_FOUR

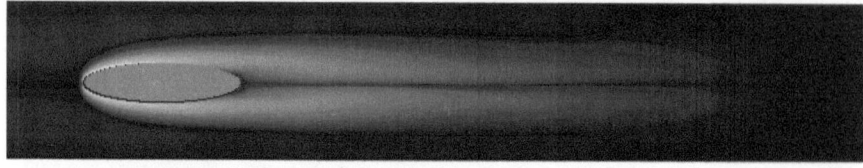

examples\nast2d\nast2d.c with option MAKE_OVAL

examples\nast2d\nast2d.c with option CONTRACTION

106

also by D. James Benton

3D Articulation: Using OpenGL, ISBN-9798596362480, Amazon, 2021 (book 3 in the 3D series).

3D Models in Motion Using OpenGL, ISBN-9798652987701, Amazon, 2020 (book 2 in the 3D series.

3D Rendering in Windows: How to display three-dimensional objects in Windows with and without OpenGL, ISBN-9781520339610, Amazon, 2016 (book 1 in the 3D series).

A Synergy of Short Stories: The whole may be greater than the sum of the parts, ISBN-9781520340319, Amazon, 2016.

Azeotropes: Behavior and Application, ISBN-9798609748997, Amazon, 2020.

bat-Elohim: Book 3 in the Little Star Trilogy, ISBN-9781686148682, Amazon, 2019.

Boilers: Performance and Testing, ISBN: 9798789062517, Amazon 2021.

Combined 3D Rendering Series: 3D Rendering in Windows®, 3D Models in Motion, and 3D Articulation, ISBN-9798484417032, Amazon, 2021.

Complex Variables: Practical Applications, ISBN-9781794250437, Amazon, 2019.

Compression & Encryption: Algorithms & Software, ISBN-9781081008826, Amazon, 2019.

Computer Simulation of Power Systems: Programming Strategies and Practical Examples, ISBN-9781696218184, Amazon, 2019.

Contaminant Transport: A Numerical Approach, ISBN-9798461733216, Amazon, 2021.

CPUnleashed! Tapping Processor Speed, ISBN-9798421420361, Amazon, 2022.

Curve-Fitting: The Science and Art of Approximation, ISBN-9781520339542, Amazon, 2016.

Death by Tie: It was the best of ties. It was the worst of ties. It's what got him killed., ISBN-9798398745931, Amazon, 2023.

Differential Equations: Numerical Methods for Solving, ISBN-9781983004162, Amazon, 2018.

Equations of State: A Graphical Comparison, ISBN-9798843139520, Amazon, 2022.

Evaporative Cooling: The Science of Beating the Heat, ISBN-9781520913346, Amazon, 2017.

Forecasting: Extrapolation and Projection, ISBN-9798394019494, Amazon 2023.

Heat Engines: Thermodynamics, Cycles, & Performance Curves, ISBN-9798486886836, Amazon, 2021.

Heat Exchangers: Performance Prediction & Evaluation, ISBN-9781973589327, Amazon, 2017.

Heat Recovery Steam Generators: Thermal Design and Testing, ISBN-9781691029365, Amazon, 2019.

Heat Transfer: Heat Exchangers, Heat Recovery Steam Generators, & Cooling Towers, ISBN-9798487417831, Amazon, 2021.

Heat Transfer Examples: Practical Problems Solved, ISBN-9798390610763, Amazon, 2023.

The Kick-Start Murders: Visualize revenge, ISBN-9798759083375, Amazon, 2021.

Jamie2: Innocence is easily lost and cannot be restored, ISBN-9781520339375, Amazon, 2016-18.

Kyle Cooper Mysteries: Kick Start, Monte Carlo, and Waterfront Murders, ISBN-9798829365943, Amazon, 2022.

The Last Seraph: Sequel to Little Star, ISBN-9781726802253, Amazon, 2018.

Little Star: God doesn't do things the way we expect Him to. He's better than that! ISBN-9781520338903, Amazon, 2015-17.

Living Math: Seeing mathematics in every day life (and appreciating it more too), ISBN-9781520336992, Amazon, 2016.

Lost Cause: If only history could be changed..., ISBN-9781521173770, Amazon, 2017.

Mass Transfer: Diffusion & Convection, ISBN-9798702403106, Amazon, 2021.

Mill Town Destiny: The Hand of Providence brought them together to rescue the mill, the town, and each other, ISBN-9781520864679, Amazon, 2017.

Monte Carlo Murders: Who Killed Who and Why, ISBN-9798829341848, Amazon, 2022.

Monte Carlo Simulation: The Art of Random Process Characterization, ISBN-9781980577874, Amazon, 2018.

Nonlinear Equations: Numerical Methods for Solving, ISBN-9781717767318, Amazon, 2018.

Numerical Calculus: Differentiation and Integration, ISBN-9781980680901, Amazon, 2018.

Numerical Methods: Nonlinear Equations, Numerical Calculus, & Differential Equations, ISBN-9798486246845, Amazon, 2021.

Orthogonal Functions: The Many Uses of, ISBN-9781719876162, Amazon, 2018.

Overwhelming Evidence: A Pilgrimage, ISBN-9798515642211, Amazon, 2021.

Particle Tracking: Computational Strategies and Diverse Examples, ISBN-9781692512651, Amazon, 2019.

Plumes: Delineation & Transport, ISBN-9781702292771, Amazon, 2019.

Power Plant Performance Curves: for Testing and Dispatch, ISBN-9798640192698, Amazon, 2020.

Practical Linear Algebra: Principles & Software, ISBN-9798860910584, Amazon, 2023.

Props, Fans, & Pumps: Design & Performance, ISBN-9798645391195, Amazon, 2020.

Remediation: Contaminant Transport, Particle Tracking, & Plumes, ISBN-9798485651190, Amazon, 2021.

ROFL: Rolling on the Floor Laughing, ISBN-9781973300007, Amazon, 2017.

Seminole Rain: You don't choose destiny. It chooses you, ISBN-9798668502196, Amazon, 2020.

Septillionth: 1 in 10^{24}, ISBN-9798410762472, Amazon, 2022.

Software Development: Targeted Applications, ISBN-9798850653989, Amazon, 2023.

Software Recipes: Proven Tools, ISBN-9798815229556, Amazon, 2022.

Steam 2020: to 150 GPa and 6000 K, ISBN-9798634643830, Amazon, 2020.

Thermochemical Reactions: Numerical Solutions, ISBN-9781073417872, Amazon, 2019.

Thermodynamic and Transport Properties of Fluids, ISBN-9781092120845, Amazon, 2019.

Thermodynamic Cycles: Effective Modeling Strategies for Software Development, ISBN-9781070934372, Amazon, 2019.

Thermodynamics - Theory & Practice: The science of energy and power, ISBN-9781520339795, Amazon, 2016.

Version-Independent Programming: Code Development Guidelines for the Windows® Operating System, ISBN-9781520339146, Amazon, 2016.

The Waterfront Murders: As you sow, so shall you reap, ISBN-9798611314500, Amazon, 2020.

Weather Data: Where To Get It and How To Process It, ISBN-9798868037894, Amazon, 2023.